Praise for Art

Smart C

"This book not only helps you unders the tools to help you change your behavior. As you work through the innovative exercises in this book, you will create new habits at home, at school, or at work—and be on your way to a healthier and happier you. This book will have lasting impact."

—Michael F. Roizen, MD, anesthesiologist, internist,
and chief wellness officer of the Cleveland Clinic,
author of *RealAge: Are You as Young as You Can Be?*,
and coauthor of the *You* series

"Technology promises to make our lives easier and more efficient, but too often it has just the opposite effect—we feel rushed, harried, and 'always on,' and we take refuge in bad habits like procrastination. If this sounds like you, you need to try *Smart Change*. It's based on the science of motivation and designed to tap into our habit-forming circuits and use them to create a more satisfying life."

—Daniel H. Pink,
author of *To Sell Is Human* and *Drive*

"Art Markman's smart, engaging, quirky book will teach you how to do more of the things that make you healthy, wealthy, and happy, and less of the things that don't. *Smart Change* is an important book that's also a delight to read."

—Adam Alter, assistant professor of marketing
and psychology at New York University's Stern School of Business
and author of *Drunk Tank Pink*

continued . . .

"A smart book for all of us who have tried to change our behavior and failed . . . Markman leverages the latest scientific research on habits and motivations to offer a practical guide to successful behavior change. In an engaging, easy-to-follow style, the book explains how habits thwart our best intentions but also help us realize our goals. As Markman explains, changing our behavior often involves changing our environments so that it's easy to repeat desired actions until they become the habitual, default response."

—Wendy Wood, PhD, provost professor of psychology and business and vice dean for social sciences at the University of Southern California

"*Smart Change* presents the science and practice of what might be the toughest job that people take on—to change themselves. Five clear and actionable steps outline the most effective ways that people can change their habits, minds, and bodies. This book's combination of cutting-edge research and narrative storytelling gives insights that are sure to stick."

—Kathleen D. Vohs, PhD, University of Minnesota and editor of *Handbook of Self-Regulation: Research, Theory, and Applications*

"Change is hard, but it doesn't have to be. In this insightful book, Art Markman gets to the root causes of our habits and explains what it takes to reshape them and make lasting, positive changes in our lives."

—David Burkus, author of *The Myths of Creativity*

"*Smart Change* achieves a rare feat—integrating the very latest science in an accessible, practical framework people can use to bring about lasting behavior change in themselves and others."

—David Neal, PhD, psychologist and behavior change expert, and founding partner at Empirica Research

smart change

Five Tools to Create New and Sustainable Habits in Yourself and Others

ART MARKMAN, PHD

A PERIGEE BOOK

A PERIGEE BOOK
Published by the Penguin Group
Penguin Group (USA) LLC
375 Hudson Street, New York, New York 10014

USA • Canada • UK • Ireland • Australia • New Zealand • India • South Africa • China

penguin.com

A Penguin Random House Company

Perigee trade paperback ISBN: 978-0-399-16412-5

The Library of Congress has cataloged the Perigee hardcover edition as follows:

Markman, Arthur B.
Smart change : five tools to create new and sustainable habits in yourself and others /
Art Markman, PhD.—First edition.
pages cm
Includes bibliographical references and index.
ISBN 978-0-399-16411-8 (hardback)
1. Habit. 2. Change (Psychology) 3. Habit breaking. 4. Behavior modification. I. Title.
BF335.M26 2014
152.3'3—dc232013034778

PUBLISHING HISTORY
Perigee hardcover edition / January 2014
Perigee trade paperback edition / January 2015

PRINTED IN THE UNITED STATES OF AMERICA

10 9 8 7 6 5 4 3 2 1

Text design by Kristin del Rosario

To Dedre Gentner and Doug Medin—

the best mentors anyone could hope to have.

CONTENTS

FOREWORD

Few books and little of the self-help advice I hear can create lasting change. That's because behavior change is hard. I should know: I have spent most of my career trying to encourage people to make healthier choices. Everyone knows the general rules for being healthy: Avoid tobacco and secondhand smoke. Walk ten thousand steps a day. Do cardio and resistance exercises regularly. Eat only foods that are good for you—and not too much of them. Meditate daily to manage stress. Find time to relax. Get regular sleep.

The problem is not a lack of knowledge about what to do. The problem is that the alternatives are too tempting: Not taking the extra steps to go to the side of the building where you'll more likely avoid secondhand smoke as you enter. Ordering a juicy hamburger with fries. Having an extra serving of dessert. Parking closer instead of farther away. Sitting while you talk on the phone—and not using a wired earpiece. Not getting the flu vaccine. Staying up just one more hour to watch a movie. None of these actions on their own seems like a problem. But they add up to a problem in the long run. Or sometimes life just gets in the way.

At the Cleveland Clinic, we have spent the past decade encouraging people to change their behaviors. It hasn't been easy. We have had to take some radical actions to help our caregivers/

employees do what they need to do to get healthier. And it has worked. Over the past several years, we have spent a lot less money each year on employee healthcare because the people who work for us have *chosen* to become healthier.

If you want to change your behavior, it would be great if a big company stepped in and helped you to reach your goals. However, all too rarely does that happen.

But you are not out of luck. *Smart Change* is on your side.

I think Art Markman is the top cognitive scientist in his field. He has been doing cutting-edge research for two decades, examining thinking and motivation. And for the last ten years, he has been taking this work beyond the scientific community to help people learn about themselves and use science to live better.

In this book, Art unlocks the mysteries of motivation. He explores why your attempts to change your behavior in the past have failed by teaching you about how your brain really works. More important, he teaches you how to succeed.

He explains why the changes that we've seen at the Cleveland Clinic were so effective, and he enables you to make the same kinds of transformations in your life, no matter what new habits you are trying to create. Art knows that making changes only begins with your commitment to do something new. You'll need to formulate a workable plan, to rearrange your environment, and to engage with people around you if you really want to realize a new *you*.

These same tools can also be used to help people around you change their behaviors, and this book will show you how. *Smart Change* is a great instruction manual for influencing others.

This book not only helps you understand yourself but also provides the tools to help you change your behavior. As you work through the innovative exercises in this book, you will create new habits at home, at school, or at work—and be on your way to a healthier and happier you. This book will have lasting impact.

—Dr. Michael F. Roizen is an anesthesiologist and internist and the Chief Wellness Officer of the Cleveland Clinic. He is the author of *RealAge: Are You as Young as You Can Be?* and coauthor of the successful series of *You* books.

ONE

the problem of behavior change

Why Is Behavior Change Hard?

Systematic Goal Failure Signals the Need for Change

Follow the Path to Smart Change

MIKE ROIZEN IS A PHYSICIAN ON A MISSION. HE knows that the formula for staying healthy isn't that complicated. You do need a little luck, but if you eat good food, exercise regularly, avoid cigarettes and drugs, and keep your stress level down, you're putting yourself in a good position to live a long and healthy life.

Unfortunately, although the formula is straightforward, sticking to it is not easy. And nobody is more aware of that fact than Dr. Roizen. He's the wellness director at the Cleveland Clinic. He knows that when people take care of themselves, they feel better, live longer, and help keep healthcare costs down. Healthy people don't need lots of doctor's visits, hospital stays, expensive tests, and long-term medications. In Roizen's opinion, businesses need to care more about their employees' wellness before healthcare costs swamp their profits. So good health is not just important for individuals, it is crucial for the world's economy.

And yet, people consistently fail to take care of themselves. They eat too much, exercise too little, drink too much alcohol, smoke, and take on too much stress. Although advances in medical technology are helping people live longer, many of the conditions people are treated for are preventable. If people focus their current behavior on factors that promote long-term health, they're much less likely to suffer the diseases that bring them into contact with all of that medical technology.

What Dr. Roizen also knows is that people can change. This isn't just a belief. He has good evidence. In 2005, the Cleveland Clinic began a new round of wellness programs for their employees. From 2005 to 2012, employees of the clinic have lost a staggering three hundred thousand pounds collectively. Smoking rates have gone down to under 7 percent, which is less than half of the national average. And in this era of rising healthcare spending, the Cleveland Clinic went almost four years without any increase in the cost of healthcare for their employees.

How did they do it?

There are two ways to answer this question.

The first is to point to a large number of programs that the Cleveland Clinic created to enhance employee wellness. In 2005, the clinic banned smoking from their campuses and started to offer smoking-cessation classes to employees who smoke. In 2008, they began offering free yoga classes to employees. In 2008, they gave out free gym memberships to employees. In 2010, they removed sugary drinks from the vending machines and cafeterias at the center. Along the way, they also sponsored farmers' markets and bought the unsold food farmers brought to the markets in order to make it easier for employees and people in the surrounding community to get fresh produce. They created a mentor program so people trying to learn to eat better and exercise more could get feedback from others about their progress. The Cleveland Clinic invested in their employees' health, and that investment paid off.

The second is to explore why these particular programs worked. After all, almost anyone will tell you they should eat a balanced diet and avoid sugary foods, but they don't do it. Every year in January, the membership rolls at gyms across the country swell as people decide that this is the year to get in shape. By February, though, only the regulars at the gym are still there. People know that smoking is dangerous, but they do it anyhow. And if they somehow miss that valuable information, it is printed in bold type on every pack of cigarettes.

That means that it is not enough just to educate people about the right thing to do. Knowledge alone does not cause people to change their behavior. And even a commitment to change is not enough. All of those people who plunk down their hard-earned cash on a gym membership are showing a commitment to improve their fitness, but they just don't keep it up.

The combination of programs the Cleveland Clinic put together does more than just promote changes in lifestyle. It gets at the heart of the system that causes people to maintain their behavior in the first place. In that way, Mike Roizen has helped people engage in Smart Change.

For you to engage in Smart Change, first you have to understand why you act the way you do. In particular, it is crucial to recognize that most of our psychological mechanisms are devoted to maintaining established behaviors rather than changing them.

Though you may not believe it, you are exquisitely tuned to achieve your goals—whether you consider them to be positive or negative. The brain system that drives your behavior so effectively is called your motivational system.

The problem you face when changing your behavior is that every behavior you want to change now is one that your brain is trying hard to maintain. So successful behavior change requires short-circuiting the mechanisms that will keep you doing what you did in the past and then reprogramming your motivational system to do what you want to do.

The second step to Smart Change is to develop a set of strategies that will support your attempts to prevent the old behavior from happening and maintain the new behavior long enough for it to become your routine.

In this book, I introduce you to the science of the motivational system that determines why you act the way you do. This tour highlights five pressure points that provide targets for changing your behavior. For each of these pressure points, there is a corresponding toolbox stocked with strategies for changing your behavior.

As you will see, one reason changing your behavior can be so hard is that no single approach is likely to be enough to change your behavior. Instead, you may need to address all five pressure points before you succeed.

Chances are this is not the first book or article on changing your behavior that you have ever read. So why will this approach help you where others have failed? The tools are all based in our best understanding of the forces that maintain your habits. The tools themselves are all specific things you can do to maximize your ability to change your behavior. For the tools to be effective, though, you are going to have to do more than just read this book. You are going to have to do some real work.

To kick things off, I want to talk a bit more about two aspects of the way your brain is organized that make it hard to change behaviors once they have been developed: *habits* and the *power of now.*

Habits Make Behavior Change Hard

Here is a startling fact: Your brain is designed to spend as little time thinking as possible. That seems like a strange thing to say. After all, any five-year-old can tell you that people think with their brains. And our ability to think, reason, and behave flexibly is truly remarkable. Most people point to our capacity for thought as the factor that truly separates humans from other animals.

We prize our thinking abilities. Many of the people whom society admires are thinkers. The death of Steve Jobs reverberated through the online community because of his ability to think about electronic devices in new ways that transformed people's lives. We look up to film directors like Christopher Nolan and James Cameron because they can bring novel visions to the screen. Over half a century after his passing, people are still fascinated with Einstein because he ushered in a new and strange era in physics.

All of those achievements are a testimony to the tremendous flexibility of the human brain. But that flexibility comes at a cost.

Brains are very expensive to operate. Your brain weighs about three pounds. That means it is 2 to 3 percent of your body weight. But it takes up a lot of resources. It uses 20 to 25 percent of the calories you burn each day and requires a lot of oxygen and blood flow to keep it running.

Brains burn through a lot of energy because the active cells in your brain—called neurons—work hard to generate the electrical signals that carry information. Neurons are constantly manufacturing chemicals called neurotransmitters, which help pass signals from one neuron to another. All of this electrical and chemical activity requires a lot of energy to maintain.

An interesting thing about the brain is that it requires about the same amount of energy no matter what it is doing. When you are "thinking hard" (whatever that means) your brain is using about the same amount of energy it uses when you are engaged in well-practiced routines. For this reason, the brain wants to spend the least amount of time necessary thinking about something.

Because brains are expensive in terms of energy consumption, most species have fairly small ones. For animals to survive with small brains, the majority of their behaviors are instincts or habits. I got to see this firsthand several years ago, when a baby deer was born in my backyard. I don't live in the country; Austin, Texas, just has a lot of deer wandering around residential neighborhoods because we have effectively eliminated all of their natural predators. Within minutes, the newborn fawn could stand, and by the end of the day it was walking around. A day later, it had wandered off with its mother.

A deer comes equipped by nature with a variety of complex skills as standard equipment. Because these abilities are wired in, the animals are able to get around with fairly small brains. A typical deer has a brain that weighs about a half pound (which is far less than 1 percent of its body weight). So compared to people, a deer is not using a tremendous amount of energy to run its brain.

Of course, deer are also not particularly smart or flexible. All of those behaviors that allow a deer to succeed in the forest work less well when confronted with a suburban neighborhood full of cars and houses. A deer spooked by a barking dog will run, even if that dash takes it into the path of oncoming traffic on a busy street. No matter how often a particular deer is nearly hit by cars (and no matter how often they see other deer that are not so lucky), they do not adapt to this environment. Instead, drivers have to learn to watch out for the deer.

Species with bigger brains (relative to the size of their bodies) are able to act more flexibly than are species with smaller brains. That means they can overcome their instincts, learn to handle changes in their environment, and plan sequences of actions.

Humans, of course, are incredibly flexible in their behavior. The first humans lived in a world without modern technology and developed tools from rocks, plants, and animals. Humans of each era, however, are able to learn the technology of the day and to take that cultural complexity as a starting point. This flexibility has two costs, and both involve time.

Unlike the baby deer in my backyard, children do not stand up the day they are born. Or even the day after. It takes months before they can stand up on their own, and months after that to start walking. Most children cannot feed themselves for their first few years of life, and it may take longer before they can prepare their own food. In the modern world, most children don't even leave home until they are at least eighteen, and even then they are typically continuing their education. It is common for people to be in their twenties before they are expected to make a contribution to society.

So, clearly, it takes years of programming for humans to develop enough knowledge about their environment to act effectively in the world. Even if the world changes substantially during the course of our lives, we quickly adapt to it. Anyone born before 1970, for example, would have lived years before encountering a computer. Yet many of them now use email regularly.

The second cost in time is directly related to this flexibility in behavior. Because you are able to pursue so many different courses of action at any given time, you have to select the behavior you are going to perform, which also takes time.

To see what I mean, consider a simple trip to a new supermarket. The first time you go to that store, you have to rely on only your general knowledge about stores. The layout is unfamiliar. As you walk through the produce aisles, you are learning about the quality of the fruits and vegetables they have in stock. You have to search to find the items you want to buy. For each of those purchases, you may also have to get a sense of the variety of brands that the store carries. That trip to the store seems to take forever because every aspect of it requires real thought.

You cannot afford to spend that much time on all of the things you do. Literally. Benjamin Franklin may have said, "Time is money," but time is also energy. Your brain wants to minimize the amount of time you spend thinking about anything to make sure the energy cost of thinking does not exceed the value of what you are thinking about.

In any situation, then, your brain has to resolve a trade-off between effort and accuracy. Obviously, you want to do things well. You could run through a new supermarket and throw things into your cart at random. That would certainly get you out of there in a hurry. But you probably wouldn't get what you need. Then you would have to make another trip, which would waste more time and energy. The trick, then, is to spend exactly as much time as you need to get what you want.

A key way that your brain helps you deal with the trade-off between effort and accuracy is by creating *habits.* Whenever you do something successful, mechanisms in your brain relate the action you performed to the situation in which you performed it. That way, when the situation comes up again, your brain can suggest that you perform the same action.

That's why the second time that you walk into the same supermarket, the shopping trip goes much faster. You have a better sense of the store layout. You know about the quality of the produce. You remember where to look for your favorite brand of tomato sauce. You are able to get through the store faster because you don't have to think as much about what to do. You just *know.*

By your tenth trip to that same store, you are on autopilot for most of the time. You can chat away on your cell phone or think about something more pleasant while you walk quickly through the aisles grabbing your groceries. Sure, there might be one or two things you need to compare or a new product that you want to consider buying. But most of the trip is just doing what you did last time. As your habit develops, your behavior gets less flexible. You buy fewer things you did not anticipate getting. You often choose the same brands of the products you buy.

Your habits have made life better. The annoying shopping trip is now done quickly and accurately. As an added bonus, the trip also feels more comfortable than that first venture into the supermarket.

As wonderful as these habits can be, they're also one big reason why it's so hard to change your behavior. Habits develop because the actions you have taken have worked in the past. Your brain wants you to do successful things quickly, so it serves up those behaviors again and again hoping you will be as successful in the future as you were in the past.

If a behavior that worked in the past continues to be successful, your motivational system will continue to promote that habit. If the environment changes, though, and that behavior no longer leads to success, then you generally change that behavior fairly easily.

Suppose that the grocery store decides to reorganize the layout of the wall of tomato sauce. You get to the tomato sauce aisle and notice that it has been organized differently. Perhaps your habit is to look for jars that are at about eye level and slightly to the left. You look there, and the brand you want is no longer in that location. You look around some more and find that now that brand has been moved to a low shelf on the right. You might mistakenly look up and to the left a couple more times on later shopping trips, but soon you will be looking down and to the right. The old behavior is gone, and the new one replaces it because the old behavior no longer helps you reach your goals.

What really makes behavior change difficult, though, is that sometimes you want to change your behavior even though the old behavior is technically still successful. When you want to diet, for example, your old eating behaviors are still working as far as your motivational system is concerned. You want to eat, and you eat. The problem is that you have some new goals. You may want to lose weight or to feel better. Perhaps your doctor has recommended that you get in shape. So you are now trying to stop satisfying one goal and start satisfying another instead.

In that case, the habits are no longer directing you toward the action you want to take. You have to find a way to stop yourself from doing what you did last time. And that means fighting your brain's tendency to minimize energy.

Unfortunately, that is only part of the problem.

I Want It Now!

One of the most difficult things for people (and other animals for that matter) is to pass up something desirable right now in favor of something else in the future. A piece of cake now has a stronger pull on our actions than having a beach-ready body by next summer. An afternoon playing video games feels like a better way to spend an afternoon than studying for an exam that won't happen for another week. Checking a few emails at work feels more productive than putting in effort on a big long-term project that won't be finished for months.

You're wired to want to do what feels good in the short term.

This is not a new observation, of course. The Judeo-Christian religions put this aspect of behavior front and center in their ultimate top ten list—the Ten Commandments. A few of those commandments are about keeping people focused on one religion rather than another. The rest, though, have to do with the trade-off between doing what feels best right now as opposed to doing what is best in the long run.

That shiny and valuable thing that belongs to someone else? Don't steal it.

That person who has gotten you so very angry right now? Don't kill him.

Your parents, who have just annoyed you for the millionth time? Give them respect.

Those desirable people you meet who are married to someone else? Don't sleep with them.

The Bible did not choose to put these commandments into their top ten because they were easy to follow. They put them there because it is extremely difficult to overcome what feels right in the here and now to do what you know is right in the long run.

A beautiful experimental method for looking at the difficulty of overcoming temptations was developed by Walter Mischel in the late 1950s to study how children deal with temptation. In his studies, an experimenter asked four-year-old children to sit in a room and showed them a small tempting object like a marshmallow or a cookie. Then the experimenter placed the object in front of the child next to a bell and said that he had to leave the room for a while. If the child was able to wait until the experimenter got back, then the child would get a larger reward (like two marshmallows or cookies). However, the child could ring the bell at any time and just take what was on the plate. If the child was able to wait fifteen minutes, then the experimenter would come back in the room and give the child the bigger reward.

In a typical experiment, the average child was able to go only about ten minutes before ringing the bell. Figured into that average are the many children who rang the bell soon after the experimenter left the room and others who were able to wait the full fifteen minutes. Even for those children who are successful at waiting, it can be a difficult experience. In fact, there are a number of popular videos on YouTube of children in these experiments agonizing over the marshmallows in front of them. It is funny to see the children struggle to wait for the bigger reward. But the videos also illustrate how hard it can be to overcome temptation.

You might think that this study says something primarily about children and the difficulty that they have trading off on something pleasant for a larger reward in the future. However, research by Mischel, his student Yuichi Shoda, and their colleagues has followed a group of children for years after their initial experience with the marshmallow task. Children who were able to hold out for an extra marshmallow at the age of four were also likely to stick with tasks as adolescents, to plan ahead, and to do well in school.

In the next chapter, we'll see that people do get better at this with age because the brain matures, but even for adults having something right now exerts a more powerful influence on behavior than waiting for something that might be better in the future. And as we get older, the things that we can have right now are more desirable than just a single marshmallow. So the temptations get bigger.

PUTTING THE INFORMATION TOGETHER, THEN, THERE ARE two big hurdles to changing behavior effectively. Your brain wants to conserve energy, so it prefers that you do what you did last time whenever your past actions have led to success. Even when you might want to change your behavior, though, there is a powerful tendency to want things that are good right now rather than waiting for something even better later. If you are trying to change your behavior so you now forego something that would be pleasant to have right now in favor of what is best for you in the future, you are going to find it difficult.

Putting Behavior Change in Context

This is a book on behavior change, so the focus is on undesirable behaviors and ways to turn those into desirable ones. Before launching into that discussion, it's worth remembering that most of you do a reasonably good job of negotiating the world successfully most of the time.

There are three main reasons you are generally successful at what you set out to do.

First, your physiology helps you keep your body reasonably healthy. You feel hunger when you need to eat and have not eaten (and are not engaging in your routines to find food). If you don't sleep enough, you get tired and your body tells you to slow down. When you do things that might damage your body, you feel pain. In these ways, your biology has given you mechanisms that help you treat yourself reasonably well. You can short-circuit these mechanisms sometimes by taking drugs or eating foods that fool your body into thinking you are doing good things for it, but most of the time, your body manages to keep you on track to stay fairly healthy.

Second, your culture helps you achieve many goals by creating structures that make it easier to perform many actions than it would be if you worked alone. Your desire to communicate with others is supported by the various networks that the culture has set up, including the postal system, phone lines, and the Internet. Your aim to travel is supported by roads and trains and airports and other systems that have developed to get you from place to place. Your need to learn about the way the world works is made easier by schools that you can attend from childhood onward. Modern culture also allows people to set up businesses to provide products and services that support your attempts to achieve the goals that are important to you by making use of other people's expertise.

Third, many of the habits you have developed over the years are good ones. If you got a college degree, for example, you succeeded in part because you developed a number of habits that helped you study and learn new material. You had habits to devote some parts of your day to classes and other parts to enjoyable activities. Getting a college degree is not easy, but good habits allowed you to keep moving toward graduation without having to plan out the fine details of each day.

Despite the many successes you have had in your life, though, you have failed at lots of things as well. You have started projects and never finished them. You have taken up hobbies for a few weeks and then put them aside, not because they were not interesting but because you never found a way to integrate them into your life. You have missed family events that you hoped to attend. You have missed deadlines for assignments or projects at work.

Failure itself is not a bad thing. Indeed, there are two kinds of failure that are guaranteed to happen in life, and one of them is actually good.

The good kind of failure comes from the way you learn how to make the trade-off between effort and accuracy. Effort is expensive because it takes up time and energy. In general, though, the more effort you expend the more accurate your performance. If the situation you are in is not that important, then you can probably get away with acting quickly. On the other hand, when you're making a choice or doing an activity that has very important consequences for you, it might be worth putting in the effort to ensure that you do the best job possible or make the best choice possible.

Imagine, for example, that you're experiencing some pain and are looking for a medication to help relieve that pain. Because relieving pain has important consequences for you, it is probably worth spending some time making sure you get the right pain reliever. After all, if the pain killer is not strong enough, you will still be uncomfortable. If it is too strong, though, it may have side effects like making you drowsy, which you do not want to deal with. In this case, you might call your doctor, speak to friend, check things out on the Internet, and speak to the pharmacist before making a decision.

Not every choice you make requires this degree of effort. If you get to the front of the checkout line at the pharmacy and decide that you want some candy, you probably have your pick of fifty different varieties. However, it is not crucial that you get a peak candy experience. In this case, you can probably make a choice quickly without consulting lots of other people for advice. Even if you don't get the perfect piece of candy, chances are you will get something satisfactory.

How can you figure out the right amount of effort to put into a choice? If you always make the best possible decision in every situation, then you are probably spending too much time and effort on some of the choices you are making. So the way you learn how much effort you need to put in is to experience some situations in which you put in too little effort and make a bad decision. That is, you have to fail. This kind of failure is fine because it helps you learn the trade-off between effort and accuracy.

The second kind of necessary failure arises because you have limited resources. You simply do not have enough time, money, energy, and other resources to do everything you want to do when you want to do it. A key part of succeeding in life is figuring out how to deal with limited resources to do the best job possible of achieving as many of your goals as possible.

If failure is not inherently a problem, then what is the signal you need so you change your behavior?

It turns out that there is a good way and a bad way to fail.

The good way to fail is to do so unsystematically. Consider the busy life of Natalia, a married working mother. On one day, she might miss a child's performance at school to finish up a project for a deadline at work. Another time, she might skip a meeting with a client to take another child to the doctor. She and her husband might go a month without eating any meals from restaurants to save money for a new appliance. And later, they might decide to keep a car for an extra year and to use the money for a vacation.

Each of these situations involves a failure. Natalia might want to attend all of her kids' school performances, but in this example she decided that her job took precedence and finished the work rather than going to the performance. However, over time, she is finding ways to satisfy all of her goals sometimes and to fail in an unsystematic way. Natalia doesn't really need to change her behavior. If she or her husband gets a raise at work, they might have some more resources to help them satisfy more of their goals, but there is nothing in her behavior that is holding her back.

The bad way to fail is to fail systematically. That is, to repeatedly fail at the same goal or the same task over and over. Let's examine the case of Jacob, a man in his mid-forties with a family history of heart disease. Jacob knows that he ought to cut back on the amount of fat he eats and that he should exercise more often. But every time Jacob has a salad for lunch, he ends up eating a huge dinner with extra meat and maybe even some ice cream for dessert. This seems to happen all the time to Jacob. No matter what his intentions are before the meal starts, he just can't seem to eat less fat.

And it never seems like the right time for Jacob to exercise either. Mornings, he has to drive his kids to school. His company has cut back on employees, so he is working long hours with high stress. By the weekend, he is exhausted. If he has any energy at all, he will cut the grass before taking a nap.

Jacob's failure is a problem because it is systematic. He is getting a lot done in his life, but he is simply not achieving some of his key goals. If he keeps up this pattern of eating, combined with job stress and lack of exercise, he could be in for some real health problems before too long.

Jacob needs Smart Change, and it begins when he recognizes there is a set of goals he's systematically failing to achieve. When you look at your own life, the consistent failures are the signals that if you continue to act as you have in the past you will not achieve your goals.

The Path to Smart Change

I started to play the saxophone as an adult. I had played the piano growing up and briefly played in a mediocre band in college. After that, the rest of life's responsibilities crowded out the time I had to play the piano, and what few skills I had atrophied. Eventually, I decided I wanted to play music again. I didn't want to have to relearn what I had forgotten on the piano, so I found a sax teacher, bought a horn, and started taking lessons.

I have been playing for around twelve years now, and I still take lessons. The best lessons I have are the ones in which I leave feeling like I don't really know anything about playing at all. During those lessons, my teacher has identified yet another weakness in my playing. I have to learn new skills to overcome those weaknesses to get better.

My teacher addresses my issues by first figuring out that *something* is wrong. He might find a problem with my tone on high notes. Then he identifies *what* I am doing wrong to cause the problem. Perhaps I am taking too much of the mouthpiece in my mouth so the reed is not vibrating enough and the tone is thin. Finally, he *creates a set of exercises* that will correct the problem. All of these problems reflect bad habits that need to be changed, and the purpose of the exercise is to create new habits that will improve my playing.

And that, in a nutshell, is the recipe for Smart Change. You need to identify the goals that you are systematically failing to achieve. Then you need to figure out what behaviors are causing the problem to determine what needs to be changed. Finally, you must develop a structure to support the creation of new habits to replace the ones that were causing the problem.

As simple as this recipe seems, it is clearly not easy to implement. Otherwise, you would be better at changing undesirable behaviors than you are. That is where this book comes in.

My core assumption is that the more you know about the way

your brain works, the more effectively you can engage in behavior change. So I start out with an exploration of the psychological factors that sustain behavior. That means that we need to delve into the inner workings of your motivational system. When you understand the way the motivation system works, it will be clear why your habits and your tendency to focus on what is good for you in the short term have such a big influence on the way you act.

As I mentioned earlier, this examination of motivation will identify five pressure points that you can use to change your behavior. For each of those pressure points, I present specific tools you can use to help you change your behavior.

This book is written not only to help you change your own behavior and but also to help you guide other people to change their behavior. While most of the book is built on an individual perspective, it turns out that the same tools you use to change your own habits can also be used when your goal is to influence the way other people act. (In Chapter 9, I change the focus of the advice from your own behavior to the actions of others.)

As you start thinking about changing your behavior, I recommend patience. Remember that the behaviors you want to change did not develop overnight, and they will not go away overnight. Behavior change takes time, and you are likely to fail a few times before you succeed. As I discuss in Chapter 8, you will have a hard time succeeding at making significant changes in your life if you beat yourself up after each failure.

Finally, the tools I describe here are used most effectively in combination rather than individually. One of the primary reasons Mike Roizen and his colleagues have had such great success at the

Cleveland Clinic is that their programs use a variety of approaches to change behavior rather than just one. By addressing all of the pressure points in the motivational system, you are maximizing your chances for success. To make sure that you see how these tools can work together in your life, read through the whole book before starting to change your behavior.

your smart change journal

Changing your behavior is difficult, and that means you are going to have to do some work if you are going to succeed. There is no magic incantation you can say that will allow you to wake up tomorrow and act differently. If you read this book straight through and then put it on the shelf and continue on with your life, then nothing will change.

Smart Change is an instruction book for changing your behavior. You are going to have to act on the instructions for the tools I present to work. To get started, you have to think about your behaviors and to plan for change. The change you want to make might be something in your personal life like losing weight or learning to paint. It might be something at work like picking up a new skill or working toward a promotion.

To help you on your way, you need to start a Smart Change Journal that will serve as a workbook for your efforts. You have a few options here. You can buy a shiny new notebook and get some colored pens or pencils. You can set up a document on your computer or tablet. (I have included a template with prompts on page 243. Or you can use the template for your Smart Change Journal that I have posted at smartchangebook.com. Just look under the "Smart Change" tab to download a copy.) Whether it's a physical book or a document on your computer, tablet, or smart phone, keep your Smart Change Journal in a handy place so that you can keep working on it.

Throughout the following chapters I will refer you to your Smart Change Journal with prompts to record the issues you are facing and the possible solutions you can devise. These passages are highlighted with the icon.

The Takeaways

Changing behavior is hard, but it can be done. The central problem is that your motivational system is not extremely effective at helping you achieve your goals. If you want to change your behavior, you have to overcome the efficiency of this system.

You can identify the behaviors that need to be changed by looking for systematic failures. You are guaranteed to fail sometimes, just because your resources are limited. When there are desirable goals you're unable to achieve consistently, though, it is time to figure out how to reorganize your life to allow you to succeed.

One factor that makes behavior change difficult is that your habit-learning system helps you perform actions that have been successful for you in the past without having to think about them. When you have a behavior that continues to be rewarding in some way, but you want to replace it with another behavior, you have to overcome the habit system.

A second factor that stands in the way of behavior change is that you have a strong bias to want things that are going to be pleasurable right now rather than things that will be good for you in the long run. As a result, many of your systematic goal failures are likely to involve situations in which you tend to do what is best for you right now rather than what is best for you in the future.

Finally, the recipe for Smart Change is straightforward but not easy. To change a behavior, you first have to identify the goals that you are consistently failing to achieve. Then you have to determine the actions that are getting in the way of success. Ultimately, you have to develop a course of action for behavior change that takes into account the factors that sustain behaviors.

TWO

sustaining behavior

Goals Direct Your Behavior

The Go System and the Stop System

Five Tools for Behavior Change

IN THE EARLY 1990S TWO TRACKS OF TECHNOLOGY development were on a crash course: cell phones and personal digital assistants. Advances in electronics allowed cellular phones to go from large devices usually found in cars to the portable handheld phones that are familiar today. At the same time, several companies were replacing cumbersome written calendars with small portable electronic devices called personal digital assistants (PDAs). For the first time, people could hold on to their calendar, phone directory, and to-do list on a device that would fit in their pocket.

It was only a matter of time until someone combined the two. In 1992, IBM introduced a prototype for a combined device, and in 1994 a version of this phone was put on sale. Several other companies released related products over the next few years. These pioneering devices could send and receive emails and faxes, while still serving many of the functions of PDAs, such as having an appointment book, a list of contacts, and a note pad.

The beauty of combining the PDA and the mobile phone in one device is that the two technologies enhance each other. The keyboards common on early PDAs made it easier for people to send text messages and email via their cell phone. Over time, these devices evolved to incorporate new developments. With the creation of the World Wide Web, browsers were added to the phones. Later, global positioning system (GPS) chips were also incorporated so that applications could make use of location-specific information. Touch-screen technology supported more sophisticated user interfaces.

Now, of course, smartphones are everywhere. Whenever there is a line at a checkout counter or coffee shop, at least half the people there will be doing something on their phone—whether it is sending a text, checking email, surfing the web, or playing a game. People walking down the street often have their focus turned to their phones. In a meeting, it's common for people to check their phones periodically for new messages.

Obviously, the smartphone is a useful device, but that only partially explains why people are constantly reaching into their pockets and purses to check their phones. After all, you have lots of useful devices, but few of them suck away so many minutes of your life.

People with cell phones have become slaves to habit. When they first get the phone, it is a novelty. Suddenly, they can be connected to the world on demand. Every once in a while, they check their phone. They discover that they have a new email or text or update on a social media site. Each experience helps them achieve a goal, and so they check it again . . . and again . . . and again. And before long, they are doing it unconsciously.

If you want to see the power of habit, watch people try to avoid using their cell phones. We don't allow cell phones to be used at the dinner table in our house. Yet our teenage sons often pull out their phones without thinking, despite the glare from their parents. If one of them has their phone set to vibrate when new messages come in, the situation is even worse. Once we all hear the buzz, the recipient of the new message gets increasingly agitated until finally he asks to be excused from the table. You have probably seen the same behavior in meetings or classes. For the committed smartphone user, it is hard to go an hour without checking the phone. That is why the smartphone manufactured by RIM was affectionately known as the "Crackberry."

Without knowing it, the people who combined PDAs with cell phones created a device that meshes almost perfectly with the structure of your motivational system. As a result, it exerts a powerful influence on your behavior. If you want to be able to change your behavior and to create new behaviors that will have the same influence on your life as the smartphone, you need to learn more about the way your motivational system works.

A Tale of Two Systems

I start this tour of your motivational system with your brain. A basic knowledge of what your brain is doing when you learn new behaviors and when you try to change those behaviors will help you do a better job of making changes to your life.

You may notice in this book that I talk both about the brain and the mind. Your brain is the organ that creates your mind. Sometimes, I want to highlight what the brain itself is doing (as I did when I talked about the energy consumption of the brain in the previous chapter). Other times, I want to focus on more psychological concepts like habits, goals, and attention. In those cases, I talk about the mind. That is just a handy way to separate between the physical things your brain is doing and the functions that your brain is supporting.

If you have never seen a human brain before, it looks like two boxing gloves sewn together on the pinky sides of the gloves. It is a rather strange looking pair of boxing gloves, though, because the surface has a series of humps and ridges on it. Broadly speaking, each side of the brain can be divided into four regions or lobes. The thumbs of the boxing gloves house the temporal lobe. The back of the brain (where the hand would enter the glove) is the occipital lobe. The front of the brain (where your fingers would be in the glove) is the frontal lobe. Between the frontal lobe and the occipital lobe is the parietal lobe. The frontal lobes on each side of the brain will figure most centrally into the discussion of motivation and behavior change.

The brain is made of many different kinds of cells, but the most important ones for this discussion are *neurons,* which are the cells that send electrical signals around the brain. A neuron is a cell that looks like it has some antennas around the cell body. It sends information to other cells by extending a tube (called an *axon*) that acts as a wire allowing a neuron from one area of the brain to influence neurons from another area of the brain.

Neurons use chemicals to affect each other. The antennas at the cell body sense the chemicals around it that have been left by other neurons. These chemicals cause electrical changes inside the cell. When that electrical change gets large enough, the neuron sends an electrical signal (a *spike*) down the tube. At the end of that tube is a storehouse of other chemicals. When the spike reaches that storehouse, the chemicals are dumped out of the cell so that they can affect the neurons around it. Your brain is a constant hum of electrical and chemical activity. That is why it takes so much energy to keep it operating.

The surface of the brain is gray in color. Anatomists have cleverly called this *gray matter.* There is only a thin layer of gray matter at the surface of the brain. Beneath it, the brain tissue is white. Yes, you guessed it. That is the *white matter.* If you slice a brain, you will notice that there are more layers of gray matter deep inside the brain.

The gray matter consists of lots of densely packed neurons. Most of the neurons in the gray matter have fairly short wiring tubes, so the chemicals they release affect other neurons close by. Some neurons need to send signals to more distant parts of the brain. When this happens, the axon goes deeper into the brain and goes from one region of gray matter to another. To make sure that the signal does not get lost, the neurons have insulation around them (like the insulation on wires in the walls of your house). That insulation is white, and that is the source of the white color of the white matter. Much of the white matter consists of brain wiring that goes from one region of the brain to another.

The gray matter areas deep inside the brain are structures that developed pretty early in the evolutionary history of complex animals. If you look at the brain of a rat, rabbit, or sheep, you will find similar structures in their brains.

The brains of humans differ from the brains of other animals mostly at the surface. This gray matter at the surface is called the *cortex.* The cortex in humans is much larger than the cortex in other animals. This difference is most pronounced at the frontal lobes.

When you look at the brain for the first time, it is tempting to try to figure out which parts of the brain are responsible for different aspects of behavior. Indeed, back in the nineteenth century, phrenologists believed that each area of the brain had a different function. They assumed that areas that were well developed in a particular person would be larger, and those would press on the skull and form bumps. Phrenologists claimed that they could read the bumps on people's heads to know what areas of their brain were largest and thus could determine which aspects of their character were strongest.

Starting in the 1990s, it became possible to play a high-tech version of the phrenologists' game. Magnetic resonance imaging (MRI) machines used magnetic fields to allow scientists to peer inside living brains. A fascinating observation from early work with MRI machines was that blood cells have different magnetic properties, depending on whether they are carrying oxygen. Because of this aspect of blood cells, it is possible to determine which areas of the brain are experiencing heightened blood flow. The brain directs blood flow to areas that have the most electrical activity because those regions need the most energy and oxygen. As a result, MRI can be used to find the areas of the brain that are most active when people are thinking.

Needless to say, this technique led to a lot of excitement.

Just as the phrenologists assumed that bigger bumps in the head corresponded to more developed functions of the brain, researchers in the early days of MRI technology focused on which areas of the brain were getting the most blood flow as people did different kinds of thinking. And for a while, there were lots of scientific papers identifying particular brain regions as the places where people did specific kinds of thinking. You may even have seen articles talking about how science has found the love center or the trust center of the brain.

Eventually, though, this initial excitement died down, and researchers realized the brain is not just a collection of specific units that each serves a different function. Instead, there are circuits or *systems* within the brain that influence the way people think and act. The motivational system that promotes and sustains behavior involves two circuits in particular, which I will call the *Go System* and the *Stop System*. The information that drives these systems to act comes from your *goals*.

Goals are the desired end states of behavior. All behavior is directed toward some kind of goal. Some of those goals may be driven by your biological endowment. You need to take in food to get enough energy, so you have goals relating to eating. Other goals are learned. The goal to check your email on your smartphone is learned. Goals play a critical role in the development of habits and in successful behavior change.

The Go System helps you engage in behaviors. It's the circuit that allows you to perform behaviors in your own interest and ultimately to develop and maintain habits. At first, a new behavior requires some amount of mental effort. You buy a smartphone, and you adopt the goal to look at the phone. The frontal lobes of your brain play a big role when you are first learning to perform a behavior. Your goal is held in the mind in the frontal lobes, and the Go System engages that goal. Over time, your Go System learns when and where to perform different behaviors. When you pull your smartphone out of your pocket while waiting to buy coffee, your Go System is involved. The Go System uses many of the gray matter areas that are deep in the brain—the ones we share with many other animals—to learn the relationship between places where you use your phone and the actions you perform. This aspect of the Go System works quite effectively because it has had a lot of evolutionary history to optimize it.

The Stop System engages whenever the Go System is trying to promote a behavior and there is some reason why you should not actually carry out that behavior. While sitting in an important meeting at work, for example, your Go System might encourage you to check your smartphone. Your Stop System is the one that keeps you from pulling the phone out of your pocket. The Stop System uses many areas of your brain that are in the frontal lobe. The frontal lobe is evolutionarily much newer than the areas involved in the Go System.

Let's look at these aspects of your motivational system in more detail.

Goals

What does it mean to be motivated? There are many people who bill themselves as motivational speakers, and they tend to come in one of two flavors. Some are former athletes or coaches or people who have overcome adversity. They focus their talks on what is required to overcome obstacles. Their inspirational stories give an audience energy and enthusiasm for the tasks they face in the future.

Other motivational speakers focus more on what needs to be done rather than on the energy to do it. Often they are business leaders, politicians, or members of the clergy. These speakers typically ask people to think about the path they are on. They want people to focus on what they're trying to achieve and to explore whether the actions they take day to day will allow them to reach the outcomes they desire.

It turns out that each type of speaker is addressing a different aspect of the way your motivational system deals with goals. The first type of speaker helps generate motivational energy, which psychologists call *arousal*. Without some amount of energy, you will not get anything done. Most of you have had days in which you just want to stay in bed rather than getting up to face the world. On those days, your big problem is a lack of arousal. I will talk more about arousal in Chapter 4.

But energy alone will not help you succeed.

That energy needs to be directed. And that is where *goals* come in. A goal is an end state that provides a focus for your motivational energy. The goal can either be a desirable state that you would like to bring about or an undesirable state that you want to avoid. Desirable states are things like hoping to get a new text message on your phone, pursuing a new romantic interest, or wanting to eat a delicious meal. Undesirable states are things like getting sick, being hit by a car, and losing money.

At any given moment, only a small number of the many possible goals you might have are actually influencing your behavior. Your motivational energy determines which goals are affecting you. When a goal has some amount of arousal, then it is *active*. A goal that has a small amount of motivational energy behind it is less active than a goal that is strongly aroused. The more active a goal, the more influence it has over what you are doing. A goal that has no energy at a particular time is *dormant*.

For example, most of the time the goal to check your email on your smartphone is dormant. When standing in a line at the store, this goal may receive some motivational energy. That engagement encourages behaviors that will help you satisfy that goal. It might suggest that you should pull the phone from your pocket and open your email app.

When you are performing a behavior for the first time, it requires effort and attention to achieve your goal. When you first get your smartphone, you decide to use it to check your email. You think about picking it up and going through each of the steps to use the phone. You might even be working with a friend or reading the manual, so every step of the process requires focus.

Eventually, the goal becomes associated with the situations in which you carry out that goal. At that point, the goal becomes active just because you are in the situation in which you normally pursue that goal. You may not even realize the goal has become so strongly related to the situation. The goal to check your email on your smartphone becomes associated with lots of situations, and you may reach the point at which you check your phone frequently without ever really thinking about the action at all.

The Go System

The Go System is your own personal habit-creation machine. The Go System has two parts to it. One part helps you accomplish new behaviors that require some effort. The other part tries to take behaviors you perform frequently and to turn them into habits that will allow you to act in the future without having to think about what you are doing. This habit system is always active, so that you can create habits without realizing it.

The effortful part of the Go System uses circuits in the frontal lobe of the brain that help you hold a new goal in mind while you prepare to achieve that goal. That first time you play with your new smartphone, you have to focus attention on all of your actions. If someone bothers you while you are trying to use the phone, you will lose your place in the process because you are devoting a lot of your mental energy to the task. You may also find this effortful thought frustrating and uncomfortable. It can be difficult to learn new skills.

Because your brain wants to minimize the amount of time (and energy) you devote to any behavior, the Go System is constantly on the lookout for ways to take goals you achieve often and turn them into routines you can perform quickly without having to think about them. That is, the Go System is looking for ways to create habits. The formula for developing a habit is straightforward. All you need is a *consistent mapping* and *repetition*.

A consistent mapping means that the behaviors you perform to achieve a particular goal are associated with some set of circumstances in the environment that is always the same. The environment can be the outside world but it can also be your internal world of thoughts and feelings.

Your smartphone creates a consistent mapping for you. You probably keep the phone in the same place most of the time—perhaps your pocket or a purse. You have particular situations in which you use the phone, like sitting on a bus, standing in line, or walking down the street. The phone itself also creates lots of consistent mappings. When you pick up the phone and hold it, the buttons and icons you need to press to check your messages are always in the same location.

There are consistent mappings between the environment and

your behavior across many facets of your life. Most products are designed to create this consistency to help you develop habits. The gas pedal and the brake pedal in your car are always in the same location so you can create a habit to accelerate with the gas and to stop with the brake. Musical instruments are set up so that the same key or fret allows you to produce a particular note every time you play it. Computer interfaces are developed so particular commands are available under the same menu every time you use the product.

Whenever there is a consistent mapping between the environment and a behavior, you just need to repeat that behavior several times for it to become a habit. Sometimes, you might deliberately repeat the action. In that case, you are engaging in practice. However, you don't need to practice something to create the habit. Any time you repeat an action when there is a consistent mapping, the Go System will create a habit.

Your goals also become part of the habit. When you engage in an activity repeatedly, you have a goal you are trying to achieve. Part of what your Go System is doing when it creates a habit is to engage the goal you typically pursue in that setting. Habits involve arousing a goal in a situation in which you have acted in the past and engaging specific actions that have helped you achieve that goal.

When I gave the formula for creating a habit, I said the behavior needed to be repeated "several" times. What exactly does that mean?

Habits are a type of memory. The first time you do something, you (obviously) have no memory of doing it in the past. That first day with the cell phone, you use the effortful part of the Go System to struggle through checking your email. The gray matter structures deep in your brain help you form links between the environment and the behaviors you have performed in the past. The second time you pick up the phone, though, you do have some memory of what you did before. You will still have to puzzle through some parts of using the phone, but you can also use some of the memory from the first time you used it. Those gray matter structures continue to form links between the environment and behaviors. After a while (probably somewhere around the tenth time that you check your email) you are able to pull up your memory for what you did faster than the amount of time that it takes to think about how to open your email. At that point, your Go System has put a new habit in place.

Does that mean it always takes about ten times to be able to do something by habit?

Unfortunately not.

Your habits are part of memory. When a memory is unique, then it's easy to find. When a memory is similar to lots of other memories, though, then it can be hard to pull up the right one. When you always perform a particular behavior in a unique environment, habits are easy to form. When you have to perform different behaviors in circumstances that all feel similar, then habits are hard to form.

Suppose that after you get your smartphone you load it up with many different apps. The apps each have an icon that appears on the screen of your phone, and they are arranged on a grid. The action you need to take when you want to check your email is similar to the action you need to take when you want to browse the web. And each of these actions is also similar to what you need to do when you want to waste some time smashing virtual blocks against digital pigs. Each requires finding a particular icon and pressing it to launch the desired app. Because the phone and the icons and the actions you perform are all quite similar in this case, you may have to open each app many more times (say twenty or thirty) before it is faster to pull up the right memory for an action than it is to think more carefully about what you are doing.

The worst-case scenario for developing habits is when the environment is packed with similar situations that require different actions. In those cases, it is hard to distinguish all of the memories you need to be able to retrieve the behavior you want rather than having to think about it.

Perhaps the most difficult case of habit creation is arithmetic. There are two ways for the Go System to solve a basic arithmetic problem. You can count or you can just associate the answer with the problem. The basic arithmetic facts are all quite similar. They involve lots of the same numbers combined in different ways that each requires a different answer. As a result, you spend whole years in elementary school learning the basic facts of addition, subtraction, multiplication, and division.

For a long time, this similarity will make it hard to develop a real habit. When you see the problem 2 + 4, you will pull out the answer to that problem plus answers to related problems like 2 + 3 and 4 + 5 and even 2 × 4, if you have started learning multiplication. As a result, it will be faster to count than to pull the answer out of memory. Only after a whole lot of practice will your memory be faster than counting. Only at that point do you really have a habit.

Finally, you may have noticed that in this formula for habits, I did not mention rewards. Lots of popular discussions of habits include rewards as a part of the formula for habits. They suggest that without a reward, there won't be a habit.

The problem is that those discussions confuse the mechanisms that the brain uses to form a habit with the formula you need to follow to create a habit. Those gray matter structures deep in your brain need some kind of signal to tell them when to link up the environment to a behavior. When you successfully complete a goal, there is a signal that relates to goal completion that tells the Go System it is time to learn. This signal also occurs when there's an unexpected event like getting a prize because your brain wants to learn what you were doing when something unexpected happened. Neuroscientists have called the brain activity that accompanies goal completion or unexpected prizes the *reward signal.*

The origin of this term comes from the way that experiments are done to understand the function of the brain. Much of our understanding of the way the brain works comes from experiments involving animals like rats and mice. You can teach a rat to perform a behavior and then measure what is happening in the brain as it learns. One skill researchers often teach rats is to have them press a bar whenever a light goes on. Because you can't tell a rat what to do, you give it a reward, like food, whenever it does something that you want it to do. For the rat, the food is an unexpected prize, so it learns to repeat whatever it was doing when it got the prize.

When you measure what's going on in the brain of the rat during learning, the unexpected prize creates brain activity. This brain activity is associated with learning. And because it happened every time the neuroscientists gave the rat a reward, they called it the *reward signal,* even though it is actually more general than that.

Unfortunately, the term can be confusing because it can lead people to think that habits can happen only when there is some kind of reward for the behavior. However, you can develop habits for mundane things you do in life like flipping the light switch in the bedroom. The first time you walk into the bedroom in a house, you look for the light switch and flip it on. When the light goes on, that gives the Go System the signal to learn that behavior. And after a while, you are able to reach out and flick on the light switch without thinking about it. There is nothing deeply rewarding about having the light go on, though. You achieved your goal, and the Go System used the signal to help it learn. So the Go System uses signals to tell it what to learn, but those signals need not be rewards of some kind.

How Do You Know You Have a Habit?

The next time you find yourself standing in line at a coffee shop or theater, watch what people are doing with their cell phones. They pull them out of their pocket absent-mindedly, and thumb through their email or the latest time-wasting game. Their fingers fly across the screen as they swipe through menus to find the particular app they want to use at that moment. They tap quickly as they send yet another text to a friend. Their phone use is clearly governed by habits.

There are four key signatures that people are doing something by habit.

First, they're not aware of what triggered them to start the action. When people walk into the coffee shop and stand in line, that situation is highly associated with pulling out the phone. And so they reach into their pocket or purse, grab the phone, and start texting. At that moment, they have no sense of deliberately deciding to use the phone. Instead, they start the search for the phone without consciously electing to do so.

Second, they can perform the habit while doing more effortful things at the same time. Most smartphone users, for example, have become highly practiced at using the keyboards that allow them to type text messages and emails. These keyboards are typically laid out using the standard QWERTY format that is used for most computers and (gasp) typewriters in the United States.

Because keyboards have a standard structure that does not change (a consistent mapping) and people have a lot of experience typing (repetition), they have a habit for typing. That habit allows them to think about what they want to communicate without having to focus on how to get the information into the phone.

As a result, habits are the most efficient way for people to multitask. You can type and think at the same time, because the typing is a habit. Habits are done mindlessly, so they can be performed at the same time as other behaviors that require mindful attention.

Third, habits are easily disrupted by changing the environment. Habits create a direct relationship between a state of the world and a behavior. When the world changes, the habit can no longer be performed mindlessly.

Suppose you borrowed a friend's smartphone for a few hours. As a joke, you might rearrange all of the icons for the various applications your friend has on her phone. That practical joke would end up being frustrating, because your friend would no longer be able to call up different programs mindlessly. Instead of being able to scroll effortlessly from screen to screen and then pressing an icon that's in a familiar location, she would now have to search carefully through each screen to find that icon. A simple task that took only a second or two now becomes a chore that requires a lot more time and effort.

Fourth, because you carry out such behaviors mindlessly, you often have no memory for doing them at all. You may pull your cell phone from your pocket, check your messages, and put it back in your pocket without being able to remember that you did it. That is one reason you may sometimes be worried that you've not locked a door or shut off the oven. You performed the action without thinking about it, and so you have no memory of it.

If you are trying to diagnose whether you have a habit, these four elements can help. If you are not aware of many of the factors that cause you to do something, then you are probably doing it by habit. If you are able to perform the behavior while doing other things at the same time, then it is probably habitual. If a change to the environment makes the behavior much more difficult to perform, then it probably reflects your Go System in action. Finally, if you perform a behavior and then later have little or no memory for having done it, then you may be doing it by habit.

Active Goals and Attention

When the Go System engages, it works hard to help you achieve your goals. The Go System is exquisite in the way that it takes over your thought processes.

A student in a workshop I was teaching told me a great story that illustrated this point. She was driving to work and had to mail a bill that was overdue. She got distracted while driving and forgot to turn into the driveway of the local post office, where there were mailboxes. About a half mile after passing the post office, she realized she had forgotten to mail the letter.

It was too late to turn back because that would have made her late for work. She started thinking about how she could achieve this goal, completely engaging her Go System. As she was driving along, she passed an office complex she had driven by every single day for years on her way to work. Suddenly, she noticed a mailbox in front of the complex. She quickly turned into the driveway and sent off her letter.

This student had probably passed the office complex a few thousand times. In all that time, she had not noticed the mailbox there. But when she drove by it that day, she suddenly saw it. The Go System actually influenced what she was able to notice in the environment.

The Go System helps us see things that help us achieve our goals. This ability supports *opportunistic planning*. Many studies demonstrate that when the Go System engages a goal, you start to formulate a plan to achieve that goal. To do that, you engage habits you have used in the past. Those existing habits help you find aspects of your environment that you've used before to achieve your goal.

The Go System not only influences what you see in the environment but also affects how much you like things. In particular, the Go System makes things that will help you achieve your active goal feel more valuable than things that are not related to your goal.

I did a study with my colleagues Miguel Brendl and Claude Messner to demonstrate this point. We did this experiment in Germany in 2001 at a time when there were many habitual smokers who smoked at least fifteen cigarettes a day. We approached students at the end of a long lecture class (during which the students were not allowed to smoke). Some participants stayed in the classroom, where they were not allowed to have a cigarette. Others were brought outside. When they got outside, the experimenter lit a cigarette, and the students quickly followed suit.

All of the students in this study were also given a cup of coffee. Many smokers drink coffee while they smoke and so coffee is part of the environment associated with smoking. As a result, drinking coffee activates the goal to smoke, which engages the Go System to achieve that goal. For the poor smokers still inside the classroom, the coffee increased the strength of the goal to smoke, but they were not allowed to actually satisfy that goal. For those outside who were smoking a cigarette, the coffee was an easy way to provide enough time for the nicotine to diffuse through the bloodstream and remove the need to smoke.

After drinking the coffee, participants answered a few questions that were not really relevant to the experiment. Then they were given the chance to buy tickets for a raffle. The tickets cost about twenty-five cents apiece. Participants had to reach in their pockets and pull out money to buy the tickets. (After the study, we refunded all of the money to participants, but they did not know we were going to do that.)

For half of the participants in the study, the prize for the raffle was two cartons of cigarettes (each carton had ten packs of cigarettes in it). For the other participants, the prize was enough money to buy two cartons of cigarettes.

We were interested in how many raffle tickets participants would purchase.

When the prize was cigarettes, having an active goal to smoke made that prize a little more attractive. There was a slight tendency for the participants in the classroom, who needed to smoke, to buy more tickets than those who were outside and had just had a cigarette. In other studies we have done, we have also observed that having an active goal increases the desirability of objects that will help participants achieve that goal.

When the prize was cash, an interesting thing happened. The participants who had just had a cigarette bought a reasonable number of raffle tickets. However, those people who were still in the classroom and really wanted to smoke bought almost no tickets. At that moment, the cash prize was completely uninteresting. The Go System cleared the way for people to achieve their goal to smoke a cigarette by making everything else in the world less interesting than a cigarette.

These two forces combine to help you achieve your goals. When a goal becomes active, the Go System makes you more likely to notice information in the environment that will support your habits to achieve that goal. Not only do you notice those objects but you like them.

You also ignore items in your world that are not related to your habits for achieving the goal. Your Go System helps you overcome distractions that might keep you from success. For the habitual smokers, money was not part of their habit to smoke a cigarette. So, even though money might help them to buy more cigarettes in the future, it was not going to help them smoke in that moment.

The Go System is incredibly efficient at helping you achieve you goals. It focuses you on information that will enable you to succeed, and it causes you to ignore things that will distract you from success. There is a cost to that efficiency. Once a habit becomes ingrained, it can be very hard to keep yourself from carrying out the wishes of the Go System. Even if you decide that the behavior the Go System has learned is one you no longer want to perform.

The Stop System

The Go System cannot prevent itself from suggesting a behavior to you. That is just how it works. When you have an active goal, and the environment is consistent with that goal, your Go System will drive you toward actions you have used in the past to achieve the goal. It will also increase the arousal of goals that are associated with the environment if you don't already have a goal active.

How do you prevent a behavior once the Go System has engaged a goal? That is where the Stop System comes in. As I mentioned earlier, the Stop System involves regions in the frontal lobes of your brain. This system has two jobs. First, it has to be on the lookout for the behavior that you do not want to perform. Once it recognizes that the undesirable behavior has been engaged, it has to stop that behavior from happening.

That seems fairly simple.

Unfortunately, the Stop System is much less effective than the Go System.

The Stop System requires mental effort. You have to be able to pay attention to what is going on around you to recognize that you are doing something that you want to avoid. You also have to engage effort to stop a behavior that you have begun. As a result, habits have an advantage over the Stop System because habits can be done without effort.

Lots of research suggests you have a limited amount of mental energy you can use to do effortful thinking at any given moment. Suppose you want to avoid a particular behavior that you often perform automatically. Later, you start thinking about something, and that soaks up all of your available mental energy. At that point, you may not even notice that you have engaged the habitual behavior until you have already done it.

As an example, the culmination of any graduate career is the doctoral defense, when the student gets in front of a committee of faculty and discusses his or her project and answers questions. The tradition in my lab is that when a student has a defense, I bring food to feed the committee. So the night before the defense I go shopping and put everything in my refrigerator.

The morning of the defense, I have to bring the food from home. Of course, I have a set of morning routines for leaving my house most days. My routine does not normally include bringing food for a defense. Twice, I have left the house without the food and have had to go back home to get it. Each time, the problem was the same. I started to get ready for the day and began thinking about something else, like an email I had to answer or an errand I had to run on the way to work. I continued with my habitual routine until I was already in the car and driving. Only later did I remember that I had not brought the food with me. Because my mental energy was engaged, my Stop System could not break my normal routine so I could grab the food in the refrigerator.

Even if you do manage to stop a behavior, the Go System is persistent. It uses *cravings* to remind you that you have an active goal that you want to achieve. A craving is an unpleasant feeling that is associated with some kind of need or desire. Hunger is the craving for food. Habitual smokers will experience nicotine fits when they have not smoked. Other goals can also induce cravings, though. An avid runner who is missing a workout will experience a need to run.

You might think cravings are the signal you usually have that tells you to engage in a behavior. You eat because you are hungry. You sleep because you are tired. You smoke because you crave a cigarette.

The Go System does not really work that way. You have routines and habits that surround many of the goals in your life. Your morning routine, for example, might include sitting down to have breakfast with a cup of coffee. When you wake up in the morning, you may not feel any real hunger. As long as you continue with your normal routine, including making breakfast and brewing coffee, you may never really feel hungry.

On a particularly busy day, though, you may not have the time to have your normal breakfast. You decide to grab something from a local coffee shop on your way to work. Once you get into the car and start driving, you will notice you suddenly feel hungry. When your Go System recognizes that you have not satisfied an important goal, it sends you a reminder in the form of the craving that we call hunger.

Cravings are uncomfortable. To motivate you to satisfy your goals, the cravings create a painful sensation that can be removed only by actually fulfilling the goal. That is just one more way that the Go System works to get its way.

Perhaps the biggest problem with the Stop System is that it's fragile. Stress and overexertion of the Stop System can often make it work less effectively. Psychologists Roy Baumeister, Kathleen Vohs, and their colleagues find that if you spend a lot of time controlling yourself during the day, you'll eventually find it hard to continue exercising that control and your Go System takes over. They call this overexertion of self-control *ego depletion*.

A hard day at work can make it difficult to stick to a diet. There are days at work when you just want to scream at someone. Perhaps a boss has been pushing you, and you have to avoid saying something nasty in return. A coworker might have done a poor job on a project, and you have to find a polite and constructive way to talk with him rather than lashing out at his lousy performance. After several hours of this degree of control, you leave work with the Stop System worn down. You are suffering from ego depletion.

When you return home, you are confronted with all kinds of temptations to eat, whether it is a second helping at dinner, a bowl of chips while you are watching TV, or an ice cream cone to cool off on a hot evening. Usually, you succeed in resisting such temptations, but after that difficult day at the office, the first craving has you reaching for a bowl. The next thing you know, half the bag of chips is now in your stomach.

Ultimately, then, the Stop System is a lot like the brakes of your car. The best way to keep a car from moving is not to start it in the first place. Once the car is moving forward, though, it is best to take your foot off the accelerator long before you actually need to stop. You can hit the brakes, but they are an effortful mechanism for slowing the car down, and if you are going too fast, they may not always get you stopped in time.

Why Is Behavior Change Difficult?

On the surface, it just shouldn't be so hard to change behavior. The Go System engages a goal and promotes a behavior. The Stop System is flawed, but it often works to stop an undesired behavior. So why is behavior change so difficult?

It is important to have an answer to this question because the tools for changing your behavior have to address the difficult parts of the process. I have alluded to the difficulties in the discussion so far, but it is worth putting everything in one place in case you want to look it over later.

One factor that makes behavior change difficult is the power of now that I described in Chapter 1. Simply put, goals that are near in time get more arousal than goals that are distant in time. The more active the goal, the bigger the influence it has on behavior. Consequently, you are biased against doing things that will pay off in the long run when there is some other activity you could do now to achieve a short-term goal. This competition between short-term and long-term goals is one key source of systematic goal failure.

To put your long-term goals on an equal footing with your short-term goals, you need to recast the activities that will pay off for you in the long run in terms of specific goals you can achieve on a daily basis that will ultimately lead to long-term success.

A second aspect of behavior change that makes it difficult is that your goals often compete for achievement. Even after you start performing actions that will move you toward some kind of long-term goal, the competing short-term goal has not gone away. Existing habits that are appropriate for your environment are still engaged when you return to that environment, and so they continue to influence your actions. Smart Change requires finding a way to diminish the influence of old habits on new behavior.

The third component of behavior change that causes problems is that your life is already busy. You have limited time, money, and energy. When you try to add a new behavior into your life, you have to make room for it in your schedule. You have to divert resources that were going toward one set of goals and direct them toward a new goal. As a result, you need to have a plan that will help you deal with these conflicts. Without a good plan, you will find it difficult to integrate the new actions into your existing habits.

Tools for Change

Putting all of this together, the problem of behavior change is fairly easy to describe. Once your Go System has created a habit, it wants to perform that habit any time the environment is right for it. When the habitual behavior is no longer a good one, it needs to be stopped. Unfortunately, simply engaging the Stop System is not an effective way to change behavior in the long term, because the Stop System is effortful and the Go System is persistent.

Although the problem is easy to characterize, it is not that easy to fix. The system that creates and maintains habits is resistant to change. To change your behavior, you have to influence all of the aspects of the motivational system. In particular, there are five types of tools you need to engage.

These tools are aimed at the *pressure points* in the motivational system. When I talk about pressure points, I refer to aspects of the motivational system that can respond to elements that are under your control. The five sets of tools I present look at ways to affect the goals you have, the activation of those goals, the engagement of the Go System, and the effectiveness of the Stop System.

OPTIMIZE YOUR GOALS

Changing your behavior requires developing a new set of goals and learning to associate them with the environment. There are several difficult aspects about changing your goals. First, when people think about a goal they want to achieve, they often talk about it using terms that are too general. Imagine a woman who wants to learn to program a computer to advance at work. Just setting the goal to learn to program will not help her to achieve that goal. Instead, she has to create a much more specific plan that will engage the actual situations she encounters at work. Otherwise, her busy work schedule will swallow all of her time before she has a chance to incorporate the new activity. Only then can she succeed in achieving her goal.

Second, it is crucial to set the right kinds of goals that will allow you to sustain behavior in the long run. The classic example here is losing weight. People often set a goal to lose a certain number of pounds or to get to a desired weight. They engage in a set of behaviors that help them lose weight. But they associate those actions with the goal of reaching a particular weight. Once they achieve their goal, however, there is no new goal to keep them engaging in behaviors to remain at that weight. As a result, they soon return to the habits that caused them to gain weight in the first place. That is a key reason people trying to diet often ride a roller coaster of weight gain and weight loss.

To succeed at changing behavior for the long term, it is crucial to *optimize your goals*. That means describing your goals in a way that is specific enough to allow them to be engaged. It also means setting goals that will sustain behavior in the long run. Chapter 3 explores tools for optimizing your goals.

TAME THE GO SYSTEM

Once your Go System has learned a habit, it will try to engage that behavior whenever you return to environments in which the habit applies. To change your behavior, you must minimize the influence of existing habits on your behavior and then find ways to reprogram the Go System so the environment is associated with a different set of behaviors.

This strategy is by far the most effective form of change because it involves creating behaviors that will ultimately be engaged as habits. The alternative is to continue to allow the Go System to direct behavior and then to try to prevent the undesired behavior with the Stop System. Riding the brakes in this way is doomed to failure. Chapter 4 examines the best ways to tame the Go System. There you will learn to create a good plan, to prepare yourself for obstacles, and to manage the degree of arousal that energizes your goals.

HARNESS THE STOP SYSTEM

The Stop System may be an inefficient way to govern behavior in the long run, but it plays a crucial role in behavior change. Until you can properly tame the Go System and engage a new set of habits, it is crucial that you prevent yourself reverting to your old habits.

There are several things you can do to ensure that the Stop System works as effectively as it can. One is to recognize that engaging the Stop System is effortful and that there will be times when it seems easier to give in to the Go System and continue with the habitual behavior. Second is to explore methods for changing the way you describe the current situation to create distance between that situation and your behavior. Third is to be aware that, although the Stop System can be inefficient, your beliefs about the effectiveness of the Stop System influence its power. These strategies will be described in more detail in Chapter 5.

MANAGE YOUR ENVIRONMENT

The analysis of habits makes clear that much of your behavior is driven by your environment. Habits themselves are triggered by specific aspects of the world around you. In addition, your ability to carry out habitual behaviors is determined by the objects that are available in the world.

As a result, it is valuable to manage your environment when engaging behavior change. The more you understand the aspects of your environment that trigger your habits, the better you can modify your environment to change your behavior. Managing your environment is also important for developing new behaviors. Ultimately, you want to make desirable behaviors easy to perform and undesirable ones difficult. Chapter 6 focuses on methods for using your environment to support Smart Change.

ENGAGE WITH OTHERS

There are three kinds of relationships you have with people around you: family, strangers, and neighbors. A small number of the people in your life are family. You are close to them, and you feel as though you would do almost anything for them if necessary. At the other extreme, most people in the world are strangers. You might help a stranger, but most of the time when you do something for a stranger, you expect something of equal value in return.

The most interesting relationship from the standpoint of behavior change is the neighbor relationship. Neighbors are people who are close to you but who are not really members of your family. With neighbors, you expect to help them to about the same degree that they help you, though they need not respond in kind immediately. In real neighborhoods, your neighbors might help you plant some bushes in your garden. You do not need to pay them for their kindness immediately, but they will expect you to participate in other neighborhood activities later.

More broadly, we create these kinds of neighbor relationships with many people in our lives—neighbors, friends, and colleagues at work. These relationships can be engaged to support behavior change. They can also be used to help change behavior in others. Chapter 7 explores these kinds of relationships and examines how you can use neighbor relationships to help you change your behavior.

AFTER WE EXPLORE THE FIVE TOOLS, CHAPTER 8 EXAMINES how you can put all of the tools together to change your behavior effectively. That chapter is particularly concerned with how the process of Smart Change evolves from the early phases of change, which can be quite exciting, to the more difficult periods, when you have still not achieved your goals although you have been working to make changes for a while.

Finally, Chapter 9 turns this book on its head. Sometimes, you are more interested in changing the behavior of the people around you than you are in changing your own behavior. The same principles of Smart Change can help you influence the way other people act. That chapter examines how to use the five tools to affect other people's behavior.

The Takeaways

The motivational system aims to help you achieve your goals. Your Go System uses structures deep in your brain to allow you to transition from learning a set of behaviors that you perform with effort to having the ability to carry out routine behaviors automatically by habit. Habits emerge whenever there is a consistent mapping between the environment and a behavior and that behavior is repeated. Once a habit is in place, it will be performed mindlessly whenever the environment suggests this behavior unless the Stop System is engaged to block the action. The Stop System takes effort to use and factors like fatigue and stress can impede its effectiveness.

A key element of the motivational system is the set of goals that people pursue. Behavior is goal directed. Goals vary in their level of activation, and this activation in turn determines how strongly the goal influences behavior. Goals can be activated from many sources, including the environment and other people.

Given the structure of the habit system, there are five key tools for creating lasting behavior change: (1) optimize your goals, (2) tame the Go System, (3) harness the Stop System, (4) manage your environment, and (5) engage with others. These tools will be the central focus of the rest of the book.

THREE

optimize your goals

Specific Goals Are More Effective Than General Goals

Focus on Processes Rather Than Outcomes

Pay Attention to Positive Goals

IT WILL COME AS NO SURPRISE TO YOU THAT THE United States has a weight problem. According to the National Center for Health Statistics, about one in three adults and one in six children are obese. And even those who are not obese often weigh more than is ideal.

Because of the weight problem in the United States, the concept of weight loss is almost constantly on our collective consciousness. Lists of the most common New Year's resolutions always put weight loss and getting in shape near the top. Headlines from supermarket magazines cry out with tips to help people shed pounds and flatten their tummies.

Weight loss is big business, too. Americans spend over $20 billion a year working with companies like Weight Watchers and Jenny Craig, buying diet books, taking weight-loss drugs, and having surgery. At any given moment, more than a hundred million people—about one out of every three Americans—will claim to be on a diet. Weight loss is difficult for individuals, and it is a thorny problem for organizations that are trying to influence the behavior of other people.

Despite all of this time, effort, and money, most attempts at weight loss fail. Diets are hard to keep. Even people joining weight loss groups have difficulty reducing the calories they take in. Those who do lose weight successfully regain much of that weight after their diet is finished.

Weight loss is a perfect storm that swallows up attempts at behavior change and leaves them dashed on the rocks. People's eating habits are deeply ingrained in the motivational system, and there are physiological factors that drive the need to eat that also come into play. Most important, people set the wrong kinds of goals when they try to lose weight, and so weight loss is a great domain for illustrating what goes wrong when people set the wrong goals and how you can be more effective at changing behavior by optimizing your goals. There are three elements that create this storm.

Failing to Define Your Goals Specifically

The first piece of this perfect storm is that people do not define the problem they are trying to solve in a way that is specific enough to be put into practice. Instead, they think about weight loss abstractly. That is, they start with a very broad goal like "losing some weight" or "getting in shape."

But, what does it actually mean to get in shape? To accomplish anything, it is important that you define your goal specifically enough to be able to know when you have succeeded. What would it mean to be in shape? You might believe that you'd know fitness when you see it, but thinking like that isn't really good enough to help you achieve your goal. It is hard to measure your progress against a goal that has not been clearly defined.

You might try to define your goal again and to say that you want to lose thirty pounds, to wear a size eight dress, or to be able to run a 10K race.

These goals are a good start because now you will know when you have succeeded. But you need to go still further because these goals are described too generally to put them into actual practice. What exactly do you have to do to lose thirty pounds or to fit into a smaller size of clothing? These stated goals do not include any actions you can take that will help you achieve your aim.

You need to clarify the abstract goals you start with and then identify the specific actions you can perform that will eventually lead to the long-term outcome you are hoping for.

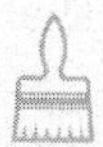

Focusing on Outcomes Instead of the Process

The second element in this perfect storm of weight loss is found in the focus people typically take on the desired outcome. Both the vague goal to get in shape and the more specific goal to lose thirty pounds are desired outcomes. That is, people who set goals like this know that at some point in the future they want to be more physically fit or to weigh less.

Several years ago, a friend's daughter was getting married. My friend told everyone that she wanted to be a "thin mother of the bride." And, somewhat miraculously within a short period of time, she succeeded. At her daughter's wedding, many people commented on how good she looked. Within a year, though, she had regained most of the lost weight. And perhaps she was happy with this. But I have to imagine that her aim was not simply to lose weight, get compliments, and then get heavy again.

Contrast this with a man I know whose doctor told him that he needed to lose weight because he was at risk for diabetes and heart disease. His own father had died in his early fifties, and so he took the warning seriously. Rather than deciding to lose weight, he set two other goals. The first was to go to the gym Mondays, Wednesday, and Fridays and to take long walks on the weekends. The second was to move to a plant-based diet. Over time, these two changes led to a remarkable transformation in his appearance. He lost an enormous amount of weight and toned up his physical condition. Perhaps more important, he is still thin and fit some ten years after that fateful visit to his doctor.

What is going on here? There are two types of goals that people can adopt. The most typical goal people pursue is an *outcome goal.* It refers to a specific state that you hope to reach in the future (like being a thin mother of the bride). The second type of goal is a *process goal* that focuses on a set of actions you can perform. As a side effect of those actions, you may achieve some desirable outcomes, but your focus is on the actions, not the outcome. Processes can go on for long after a particular outcome is reached, making them great goals for long-term life changes.

Losing Sight of Positive Goals

The third component of the perfect storm that makes weight loss—and other behaviors that involve cessation—so hard is that even after you define a specific goal, you have probably created a situation that will not engage the Go System, the circuit in the brain that allows you to develop and maintain habits. In fact, the most difficult behaviors to change involve *stopping* a behavior rather than starting a new one. Weight loss, for example, requires eating less food. Other difficult types of behavior change like quitting smoking or avoiding checking your cell phone in business meetings are similar. If you are doing something that you want to stop, then the desired state (your goal) is to be doing nothing.

However, the Go System cannot learn to do nothing. The Go System learns something new when it can find a relationship between what is happening in the environment right now and an actual behavior. The Go System is looking for ways to predict what actions it should perform in the future. It does not want to learn to predict when *not* to do something.

Trying to stop a behavior creates a *negative goal*. Negative goals do not engage the Go System; they are ultimately doomed to failure because the Go System never develops any new habits. As a result, it is up to the Stop System to keep preventing the Go System from leading you to perform an undesired behavior. Instead, when faced with the prospect of having to stop a behavior, it is crucial to create a *positive goal* that allows the Go System to learn something new. Positive goals describe actions you can perform rather than actions you are trying to avoid.

Of course, the first kind of behavior you have to learn to change is the set of behaviors around change itself. And that means we need to get more specific about the methods for optimizing your goals.

the plan for optimizing your goals

At the beginning of this book, I told you that you were going to have to do some work if you wanted to change your behavior. Now is the time to get started on that work.

SMART CHANGE JOURNAL

For the Smart Change processes to work, you need to answer a series of questions related to the discussions in this book. The best way to proceed and to monitor your progress is to create a journal in which to record your answers and observations. (I have created a document of a model journal that you can download at smartchangebook.com under the "Smart Change" tab, or you can simply start with a blank book. I provide all of the questions you need to answer in your journal throughout the book and on page 243.)

It took you a long time to create the behaviors you now want to change. You have to be willing to create a structure that will be stronger than the forces that are holding your habits in place.

Along the way, you need to resist your desire to short-circuit the process and get on with it. As you work your way through this book you may get frustrated. I don't want to belabor this point, but if you are using this book to help you change your behavior, then chances are you are dealing with something that has resisted your efforts at change in the past. If just getting on with it were the answer, you would already be done changing your behavior. I ask you only to trust in this process.

What Is the Big Picture?

To get started, think about the overall goal you want to achieve. At the top of the first page of your Smart Change Journal, write down the big picture goal you have for changing your behavior:

What Do I Want to Accomplish?

Now, take a look at that goal. How did you state it? If you are like most people, your first attempt at stating a goal is quite general. For example, the U.S. federal government maintains a list of popular New Year's resolutions. This list includes things like "drink less alcohol," "eat healthy food," "get a better education," and "save money." If you are concentrating on making changes in your work life, you may have goals like "get a promotion," "take on more responsibility," or "learn a new skill."

How do some of these items stack up to what you wrote? Notice that each of these resolutions and work goals is quite general. That is the most common way to state a new goal for changing your behavior. You focus broadly on what you would like to accomplish.

You may find you have several things you want to change. For now, that is fine. Make a separate page, and go through this exercise for each of the changes you want to make. Soon, though, I am going to ask you to pick one behavior that you want to focus on first. There is good evidence that trying to make several vastly different changes at the same time is much less effective than making one change at a time.

Why Is This Goal Important?

It is good to start out with a broad description. The management philosopher Peter Drucker distinguished between *achievements* and *contributions*. An achievement is a task that has been completed. A contribution requires accomplishing a much larger scale goal that makes a difference to an individual, an organization, or to society.

As Drucker pointed out, it is quite possible to check lots of tasks (or achievements) off a to-do list without having that set of jobs add up to anything meaningful. That is why it is important to have an abstract description of where you would like to end up. This general statement of your aim helps ensure your actions are focused on what is important to you.

Of course, Drucker also noted that focusing on making a contribution alone will not lead to success either. A contribution is ultimately made up of a sequence of completed tasks that add up to a larger whole. And that is the place where most of our attempts at behavior change fail.

For the rest of this book, I will stick with Drucker's terminology and refer to the big picture goal you are trying to achieve as your contribution. The word *contribution* often brings to mind some kind of world-changing set of accomplishments. And, indeed, you might want to transform the organization you work for or the world around you. But you can also make a contribution to your own quality of life or to your family or neighborhood.

The *specific goals* you need to develop and complete to make a contribution will be called *achievements*. Achievements are smaller-scale tasks with specific measurable end points.

Are You Really Sure This Is the Right Goal? Why?

Take a look at the big picture goal you wrote down. Does it feel like a true contribution to you? That is, are you really focused on something that you want to achieve? Is it something that is worth the time and effort you are about to put in?

Before you get started on this process, it is worth considering whether you are really focused on the right change in your behavior. Sometimes it seems obvious. If your doctor says that your health requires you to lose weight and get in shape, then that seems like the right goal to pursue.

But not every contribution you can think of is the one that is right for you. Every two years at the University of Texas, I work with a group of honors students. These students engage in a one-year research project under the supervision of a member of our faculty. I coordinate a seminar that serves as a kind of lab meeting for the students.

Most of the students start the honors seminar with the intention of going to graduate school. Some will go to medical school or law school. The rest are considering a PhD in clinical psychology or in an area of research like cognitive or social psychology.

One of the things I do with the students is to ask them to think about why they want to go to graduate school. A lot of the students have a real love of research and would like to be professors some day or to become therapists and help others. But some of the students are just stuck on a track of being in school. Research is not really a passion, but they have been in school for so long, it feels like they ought to keep going.

Over the course of the year I spend with these honors students, the ones who do not have a real passion for research or clinical work come to the realization that the contribution they are aiming for is not the right one for them. When they reach that conclusion, I ask them to start thinking about other contributions that they want to make.

I have to add that switching goals like this can be difficult to do. Often, these students have told family and friends that they are going to go to graduate school. Admitting a midcourse correction is hard sometimes, but it is still better than going through with a plan just because you don't want to admit to other people that you have changed your mind.

Before you commit to the contribution you wrote down, ask yourself why this goal is important to you. If you succeed at this big picture goal will it really get you where you want to be? Are you doing this to satisfy other people or for yourself? If you have doubts that this goal is the one you really want to achieve, then think about it for a while. Once you start on the path of Smart Change, you are going to spend a lot of time with this goal. It is worth the effort to make sure it is the right one.

If you do not think that this particular goal is the right one, you have two options. One is to see whether the initial way you stated this contribution needs a significant revision. The students in the honors program ended up replacing the goal to go to graduate school with another goal like starting a career. You may want to make a similar revision. Another alternative is to start this process over with another aspect of your behavior that you want to change. You may already have started thinking about several possible contributions, in which case you can toss this one to the side. Otherwise, start a new sheet with the heading "What do I want to accomplish?" and repeat this exercise.

Be Specific

Once you are comfortable with the contribution you want to make, you need to take this broad goal and break it down into smaller actions that will allow you to achieve it. There are three important aspects of making your goal more specific. You need to think about actions you can take, obstacles you must overcome, and how you will know you're finished.

Go to a fresh page of your Smart Change Journal. Write down your contribution at the top. Now, divide that sheet into three columns and label them "Actions," "Obstacles," and "Signs."

ACTIONS	OBSTACLES	SIGNS

ACTIONS

List actions you can take to advance you toward your goal. For now, don't worry about the order you will do these things. You will work on turning these actions into a plan in Chapter 4. Instead, focus on specific tasks you can perform that would bring you closer to your goal. You know that an action is specific when it describes when, where, and how that action will be taken.

To illustrate this process, I am going to return to the specific case of learning to play the saxophone that I mentioned in Chapter 1. I learned to play the saxophone as an adult, and so I had to figure out how to integrate the process of playing the instrument into my life as an adult. Many people looking to get new professional education go through a similar process. I run a master's program at the University of Texas called the Human Dimensions of Organizations. Students in that program have jobs so they have to find a way to integrate a rigorous education program into lives that already feel filled to capacity.

Here's how I could have used a Smart Change Journal when I decided to take up a musical instrument. I begin by writing down the contribution "learn to play the saxophone." At least to get started, that seems like a perfectly reasonable way to describe what I hope to accomplish.

After deciding to commit to this contribution, though, I quickly reach an impasse. I know I must translate my contribution into several specific actions that I can take. But I am not sure what those actions should be. So I start by setting specific goals to get more information about what is required to learn to play the sax. I know I need to own a saxophone and to have a teacher, so I pick a date on the calendar to go to a local music store and ask them how to buy a saxophone and to recommend a teacher who is willing to give lessons to adults.

As it turns out, the music store recommends that I find a teacher first, because teachers differ in their opinions about the best kind of saxophone for beginning players. They give me the phone number of a teacher. I now have a new goal to add to my list of actions: Call the teacher to talk about buying a saxophone and to schedule lessons. I also know that I need to find time in my schedule to practice.

As a part of the process of creating your own list of actions, make notes about anything you are not sure of. You will need to keep a running list of questions you need to have answered. At the back of your Smart Change Journal, allot some pages for taking notes. You can use that section for making a list of items that need more exploration.

OBSTACLES

As you begin to make goals more specific, it leads you to recognize that there are obstacles to overcome to succeed at carrying out the specific actions you have envisioned.

What are the key obstacles you face? The more you focus on particular actions related to your goal, the more obstacles that may rise up to challenge your success. These obstacles reflect the limitations in your resources, such as time, money, and personal effort. The obstacles are also a result of the variety of goals you have that take up these resources. Some of the obstacles may reflect aspects of your own psychology. There may be a contribution you want to make that requires you to achieve a goal you know you will not enjoy doing. For now, don't spend time trying to figure out how to overcome those obstacles. I will explore the process of planning for obstacles in Chapter 4.

It may feel strange to be thinking about obstacles. If you look through the self-help sections of bookstores, you come across many different approaches to changing your behavior. One popular strand of these books focuses on the role of positive thinking in success.

Perhaps the most famous early proponent of positive thinking was Norman Vincent Peale. As a minister, his trademark sermons focused on the importance of thinking positively to achieve goals. His speeches, radio programs, and writing suggested that positive thinking will give people the mental strength to achieve their goals. Peale was a deep believer both in the power of God and in the power of the subconscious mind to respond to positive thoughts. Indeed, he drew parallels between religious faith and the faith that allows people to strive toward the goals they want to achieve.

The importance of positive thinking has been carried forward in other writing about success over the years. In 2006, for example, Rhonda Byrne published *The Secret*, which also focused on the role of positive thoughts in achieving goals. This book took a mystical approach and suggested that the energy created by positive thoughts would resonate with positive outcomes in the world and would make success more likely. That is, focusing on the negative would actually make it harder to achieve your goals in life.

These messages about the importance of positive thinking are quite powerful for people who are contemplating a commitment to make a contribution. It is easy to get frustrated by failure and by the slow pace of behavior change. Failing to achieve your goals quickly can be demoralizing, and people often give up. It can be energizing to maintain a positive attitude.

If you look carefully at what makes people successful at changing behavior, though, positive thinking alone will not do it. In particular, good planning requires thinking about the obstacles that stand in the way of your goal and planning for them. It's just not enough to want something to happen. As the Boy Scout motto says, Be Prepared.

As I began to start my saxophone lessons, a number of obstacles quickly became apparent. Playing the sax required that I practice just about every day, and that meant that I had to find time in my daily routine for practice. I did not feel as if I had a lot of extra time. I didn't try to solve this problem right away. It was important just to begin to recognize where the problems were likely to be.

Being clear about the limitations in my time also required me to be realistic about my path toward the goal of learning to play the saxophone. I was not going to be able to devote hours a day to practice. I could not quit my job or give up my role as a parent just to improve as a musician. So I had to be willing to accept that it might take years before I was really able to play the instrument well.

As this example shows, there is interplay between the actions you identify and the obstacles you discover. As you find obstacles, you may choose to refine the actions you are going to take to avoid the obstacles. The new actions you select may also raise other problems that need to be addressed. And so the process continues. In addition, some of the obstacles may lead you to revise the contribution you hope to make (or perhaps the time line to reach that contribution).

You are probably reasonably good at finding obstacles. Chances are, you have talked yourself out of all kinds of changes in your behavior in the past by figuring out all the reasons you would fail. You then used the problems that could get in your way as an excuse not to try to do something new or to make a change.

Your motivation to continue can be quickly sapped when you spend time listing the many factors that could prevent you from succeeding. The things that can go wrong on your way to behavior change can seem overwhelming. However, you have to treat the obstacles you find as an opportunity to solve problems rather than as a reason to give up. There is a fine line between talking yourself out of something and preparing yourself for Smart Change. You cross that line at the moment that you decide that each obstacle is just a temporary barrier.

SIGNS

Once you think about the actions you want to take and the potential problems in your path, you need to have some way of measuring that you have succeeded. That is the purpose of the third column in your Smart Change Journal, the one labeled "Signs." What are the signs (or signals) you'll use to determine that you've accomplished your goal?

There are three aspects of signs that will help you determine you have succeeded at the specific actions you are taking as well as determining when you have made your contribution (that is, accomplished your overall goal).

1. Each of the signs needs to be to be defined as objectively as possible to remove the wiggle room you might use to convince yourself you have succeeded when you really have not. Unfortunately, one way you can enjoy the feeling of success without the benefits of attaining your goals is to redefine what it means to succeed. In many ways, this idea has gotten ingrained in our culture over the past several decades.
2. Each sign must be defined so that it can be measured, and the measurement itself must be easy so you don't have to spend too much effort trying to figure out if you have succeeded in making progress toward your contribution.
3. Signs may help you recognize when you have focused on out-

come goals rather than process goals. (I will talk more about outcome and process goals in the next section.)

All of us would like to feel as though we had reached our goals. At the end of my high school career, I put on a cap and gown and accepted my diploma. It was the first time I graduated from anything with such formalities. About ten years later, I became a parent. Just one generation later, my kids have graduated from preschool, kindergarten, middle school, and finally high school. Each of these ceremonies came along with caps and gowns and sheets of paper.

Though I am a big fan of celebrating achievements, I think a graduation should mark the completion of a contribution. By creating a series of graduations, our culture is redefining success in a way that minimizes the big picture contributions. Once you develop a habit of creating big celebrations for small moments, it gets easier to pat yourself on the back for doing something that hasn't really achieved your ultimate goal.

That means that you need to define all of your goals in ways that have specific markers of success. That way, you will know whether you have reached your goal. You might be afraid to have these specific markers, because having criteria for success also means that there are specific markers of failure. That is OK. I will return to the concept of failure again in Chapters 5 and 8.

Returning to the saxophone: When I first decided that I wanted to learn to play, there was no real sign that I'd completed this goal. There is no obvious test that divides those who play the sax from those who don't. I could have selected a general sign like "In ten years, I will like the way I sound when playing the sax." The problem with this sign is that it is highly subjective. I would be almost certain to succeed, but that would not mean that I had really learned to play the instrument.

I quickly changed my description of the contribution I wanted to make. I set the goal to play in a band within ten years. That sign was more objective. It is fairly clear what it means to be in a band, and ten years is a specific time frame.

To reach that contribution, though, I needed to set up other measurable signs of progress of the more specific actions that were part of my plan. Without some other potential signs of progress, it would be ten years before I knew whether I had progressed toward my goal. It is not possible to make any midcourse corrections without signals about where you are in the process of changing your behavior. I will return to the issue of evaluating your progress in behavior change in Chapter 8.

For each of the achievements (that is, specific actions) you have listed, think about the signs that mark their successful completion. Indeed, you may want to start refining the signs of success and completion by building them into the description of the action in the first place. If, for example, I described the action to practice the saxophone "thirty minutes a day, at least five days a week," then it is fairly clear whether I have achieved that goal for any given week.

The signs you have listed so far have been focused on the actions that you described in the left-hand column of your Smart Change Journal. You also want to make sure that you list signs that will help you mark your progress toward your contribution. Add those to your list of signs as well, but put a C next to them to help you remember that they are indications of how you are progressing toward your broader goal.

At this point, your journal should be filled with actions, obstacles, and signs. We're not done yet. These lists are like crude oil. They need to be refined considerably before they are ready to be put into practice. The next step toward optimizing your goals involves thinking about processes.

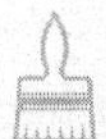

Find a Process

Speakers of the English language like to talk about end points. We tend to say things like "I bought a shirt," rather than "I went shopping for a shirt today." The first sentence focuses on the completion of a project. The second sentence focuses on the act itself. The difference between these sentences reflects what linguists call *aspect*. When you focus on the completion of the project, you are thinking about the event statically. When you talk about the process, you are thinking about the event itself, which is much more dynamic.

The aspect you use to talk about things in your lives affects your goals. Because the most natural way to express things in English is to focus on the completion of an event, that affects the way you describe your goals, which in turn influences the way you think about your goals. It feels right to say, "I want to lose weight," "I want to learn to play the saxophone," or "I want to quit smoking." It is more awkward to say, "I want to be losing weight," "I want to be learning the saxophone" or "I want to be living my life without cigarettes." These more process-focused ways of expressing yourself sound foreign.

As a result, if you're a native speaker of English, the most common way to talk about your goals is as end points. That means that you tend to phrase your goals in terms of end points as well. You talk about outcomes more often than you talk about processes. When you think about the actions you are going to take in your life that will lead you toward your bigger goals, it is a good thing to be focused on completing those actions. You ultimately want a plan that involves lots of specific actions you take in service of a larger goal. Although the focus on completion of tasks is great for actions/achievements, it is not so good for overall goals/contributions.

When you consider the big picture, it's dangerous to be focused solely on the outcome. As I mentioned at the beginning of this chapter, the problem with focusing on outcome goals is that lasting behavior change requires new *processes* in your life that will sustain themselves for the long term. When you focus your efforts on an end point, then you may not develop the specific behaviors you need to create long term change.

In Chapter 1, I illustrated the distinction between outcome goals and process goals with examples of people I knew who had physical fitness goals. In this domain, a focus on losing a particular amount of weight has two problems. First, it drives your attention toward the actions needed to lose weight rather than the actions needed to sustain a healthy weight for the long term. Second, it highlights a fixed end point (the time when the weight has been lost) rather than on creating a broad life change.

When trying to lose weight, people often engage in behaviors that are not possible to maintain in the long run. They restrict their eating in an effort to reduce significantly the number of calories they take in. They take diet medications to curb their appetite. They avoid snacks. Then they monitor their progress toward the goal of reaching their desired weight.

This combination of behaviors is not sustainable. They are all focused on the end point. What happens when you reach your desired weight? At that point, you have to change your behavior again. Suddenly, you have to develop a new set of skills for *maintaining* your weight.

Notice, though, that maintaining your weight is a long-term process that requires eating in a way that's enjoyable for years to come but is also consistent with your desired weight. It involves regular exercise. It involves a healthy overall lifestyle.

Rather than framing your contribution as an outcome goal, it is better to start your work on behavior change by focusing on a set of long-term process goals that incorporate the needed behavior change into your life from the beginning. A process goal uses the language of ongoing procedures. Rather than the outcome goal of losing a particular amount of weight, you should set the process goal of living a healthy lifestyle.

In this way, you can set up a series of behaviors like joining a gym, eating more plant-based foods, cooking at home, and adding more walking to your day. These behaviors are not focused on weight loss specifically. However, this combination of activities will lead to weight loss and better fitness as a side effect. Cooking at home, for example, allows you to control the size of the portions of food that you eat. Restaurants provide large servings, and you probably have the habit of eating all of the food on your plate. So taking control of the amount of food you eat will help you to lose weight. But you are not focused on losing weight; you are focused on learning new recipes and finding ways to cook your own meals.

The beauty of process goals is that they have no obvious end point. These goals simply become a part of the way you live your life. While you get desirable outcomes from these process goals, you avoid the single-minded attention to the outcome.

It is possible to create process goals in a variety of aspects of your life. Many people set the goal to get in a new romantic relationship. To do that, they join dating sites, go to singles mixers, and attend speed-dating events. All of these activities are designed with the goal of allowing people to develop a new relationship.

But you don't have to be focused on the outcome. Instead, set the goal to join groups of people who share your interests. Start to volunteer in your community, do more hiking, or go out to hear live music. These activities are going to bring you into contact with like-minded people, some of whom may also be looking to get involved in a relationship. Your focus is not on meeting someone, it is on the activity. You will meet interesting new people in the natural course of working toward this process goal.

The same thing holds true in a work environment. You might want to get a promotion at work, but setting an outcome goal of getting a promotion is unhelpful. It is more effective to focus on aspects of your work performance you can improve that will ultimately attract the notice of managers who can reward your efforts with more responsibility, promotions, and raises.

When I started my saxophone lessons, my entire aim was to create process goals. Finding a regular time for lessons and clearing time in my daily schedule to practice were both attempts to develop procedures that would—over time—allow me to achieve my broader goal of learning to play. And defining my contribution by hoping to play in a band reinforced the process elements of the goal, because once you join a band, you have to continue to improve your playing.

So return to your page of actions, obstacles, and signs in your Smart Change Journal. Look at the list of actions. For each action, think about the big picture contribution it's focused on achieving. Is that large-scale goal a process goal or an outcome goal?

To optimize your goals, you want to ensure you are focused on creating a set of sustainable processes for the long term. At the same time, you want to make sure that each of the specific actions you are going to take in service of that process goal is something that has a clear outcome that you can measure. In the end, you want to have outcome goals for your specific achievements that combine to lead to processes that change your life.

Find Positive Goals

Weight loss has one other aspect that makes it difficult. People tend to think of it in terms of actions to avoid rather than actions to take. Dieting is usually characterized as eating less food than before and cutting high-calorie, high-fat foods out of your routine.

In Chapter 2, I laid out the structure of the Go System and the Stop System. Because of the way these systems operate, it is difficult to change behavior when your focus is on *not* doing something. The initial mechanism you engage when trying to prevent yourself from performing an action is the Stop System. The Stop System is inefficient and can get worn down from overuse. Long-term behavior change cannot happen when your focus is exclusively on *avoiding an action*. Chapter 5 discusses ideal ways to use the Stop System, but it is important to give the Stop System as little to do as possible to be effective in behavior change.

The ultimate aim of Smart Change is to have the Go System learn a new set of habits that sustains behaviors that support your contribution. If you are able to turn desirable behaviors into habits, then you will work toward your contribution without having to think about it. Doing the right thing by habit is preferable to having to think in an effortful way about actions that will enable you to achieve your long-term goals.

The Go System only stores memories of actions. It does not store memories of inaction. After all, there are many times when you are not performing a particular action. It would overwhelm the Go System to have to learn whenever you did not do something. In the end, the aim of the Go System is to have habits that associate the environment with a behavior.

The problem with defining your goals in a negative way is that when you succeed, you don't perform any behavior at all. And if there is no behavior, then there is nothing for the Go System to learn. You simply cannot associate a context with doing nothing at all. You cannot learn to eat less. That is not an action.

There is another paradoxical effect of defining your goals negatively that comes out of research on dieting. Peter Herman and Janet Polivy point out that people who are on a diet have to spend a lot of effort thinking about the food they are eating because they have to monitor their environment for situations in which they might have to stop themselves from eating. As a result, the concept of food is always present in the minds of dieters.

If you are trying to avoid eating, it would probably be best if you did not think about food that often. So if your method of controlling your eating forces you to think about food all the time, you are putting yourself in a position in which the goal to eat is going to remain active. It is hard to distract yourself when you are constantly on the lookout for potential diet breakers. When you can no longer resist the temptation to eat, you may binge. Breaking your diet can then sap your motivation to start dieting again.

For all of these reasons, it's important to frame your goals positively rather than negatively. That means focusing your goals on actions you can take rather than actions you hope to avoid.

Go back to your Smart Change Journal. How have you defined your contribution? Is it focused positively? How about the more specific achievements? Are these actions that you can *perform* rather than things to be *avoided*? As you go through the actions on your list, make sure you highlight situations in which you are trying to avoid doing something and find ways to create processes that turn these negative goals into positive actions that can be taken.

Turn Negative Goals in Positive Actions

To return to the example of weight loss, the typical starting point is to think about going on a diet. But that is a negative goal. It requires eating less food, stopping snacks, and generally focusing on what you are *not* doing rather than what you *are* doing.

There are several positive goals that can be engaged to support weight loss. Rather than dieting, change the kinds of foods you eat. Identify the high-fat and high-calorie foods that cause you to overindulge. Replace them with other foods that are tasty, but lower in calories and fat. Find new recipes and explore new flavors. Then add other positive actions that support weight loss, like exercise. In addition to an exercise program, find new ways to add a little extra activity to your day. Commit to walking up the stairs at work or joining friends as they walk their dogs each morning. These positive goals will support actions that can be turned into habits.

Return to the goals and actions you've been working on. What can you do to keep them focused on positive actions? Think about the steps you can take that will bring you closer to your long-term goal. If you get stuck, reach out to friends and family and ask for their advice for ways to turn your goals into positive goals.

Goals and Flexibility

Like many research psychologists, I have to apply for grants to fund my studies. At the beginning of my career, I sent several applications to agencies like the National Science Foundation and the National Institute of Mental Health to seek money. A grant proposal is a little like the process of setting your goals for behavior change. In a proposal, you describe the research you want to carry out over the period of the grant (which is often three to five years). You have to lay out a contribution you want to make to the field as well as specific achievements (such as experiments) you will take to make that contribution.

I started applying for research grants when I got a job as an assistant professor at Columbia University. After several unsuccessful applications, I got a call from a program officer at the National Science Foundation telling me that they were going to fund my research. I was so excited; I jumped up from my desk to find someone with whom I could share my good news. I saw my colleague Dave Krantz in the hallway and told him that my proposal had been funded. Dave is one of the world's leading experts in measurement and a wise soul. He said to me, "That's great, Art. My only fear is that you'll do the research in your proposal."

What he meant by that was not that I should ignore the work that I had been funded to do (that wouldn't be ethical). Instead, he recognized that it is hard to have a detailed plan for research (or life for that matter) in advance. No matter how well you try to envision the future, unexpected things happen. In research, a study you do yields a surprising result that takes you off on a new path, or a student makes a comment in a lab meeting, and suddenly you are looking at your work in a whole new way. As important as it is to plan carefully, it is equally important to give yourself the flexibility to adapt to new circumstances.

The same thing is true with changing your behavior. Optimizing your goals is the first step in generating a plan for living your life in a different way. And you want to do as good a job as possible of identifying both the broad contribution you want to make as well as the narrow achievements that will get you there. But circumstances also change. And obstacles you thought would be easy to overcome can just as easily turn out to be insurmountable. Not to mention new barriers will arise that you did not anticipate.

In those cases, you need to allow yourself the flexibility to change your goals. Sometimes the change involves finding a new path toward your ultimate aim. Sometimes it involves changing the time frame in which you achieve the goal to make your plan more realistic.

The next chapter will take your goals and the actions and obstacles you identified and use them to create a concrete plan to change your behavior. As you enter into this process, it's important to realize that this plan is your best proposal for what you are going to do to change your behavior.

My worst fear is that you will try to carry out this plan without allowing yourself to adapt it over time.

The Takeaway

This chapter helped you optimize your goals. At this point, if you look at your Smart Change Journal, you should have a clear statement of the overall goal (or contribution) you want to achieve. This contribution should be stated positively rather than negatively. It probably took you several tries to find a statement of the contribution that would be most productive.

Beneath this broad contribution you listed a specific set of actions (which I am calling achievements) that you feel you can carry out. These specific goals form the steps in the plan you need to make your contribution. To maximize the quality of these specific goals, they need to have clear outcomes so you will know when you have succeeded. That means your broad contributions need to be process goals in order to make sure you incorporate behavior change into the fabric of your life. At the same time, your specific actions need to be outcome goals so you know you have succeeded.

Finally, as you go through the process of behavior change, do not feel bound by the way you stated your goals at first. Both your contribution and your achievements are tools to help you change your behavior. If you create a set of achievements you feel you cannot carry out, do not put yourself in a situation in which you cannot succeed. Change those achievements to something that gets you on the road toward your contribution but is something you can actually do. Smart Change is a process that evolves over time.

FOUR

tame the go system

Developing a Good Plan

Dealing with Obstacles

Energizing Your Goals

THE COMIC STRIP *PILED HIGHER AND DEEPER* (OR *PhD*), drawn by Jorge Cham, does an excellent job of capturing the life of a graduate student. There are a number of popular themes in the strip, such as the clever ways students waste time when they should be working, the differences in lifestyle between students and professors, and the ways that undergraduate students push their graduate teaching assistants to help them pass their classes.

One thing that comes through clearly from the comics, though, is that grad students are pretty strongly tied to the specific project they are working on as part of their thesis or dissertation. That fits with the popular image that graduate school creates a single-minded focus on research. I know I certainly believed that when I got to grad school. I expected that entering a program in cognitive psychology would allow me to take my undergraduate background in cognitive science with a heavy dose of computer science and turn that into expertise in the psychology of thinking.

What I didn't expect was the way that grad school was also going to teach me to succeed in the day-to-day process of doing science. I was lucky to have a number of great mentors. Clearly, I did learn a lot about psychology and science. The faculty at the University of Illinois gave me a broad perspective on how to look at the relevant existing studies and to find ways to add new research to the field. Looking back, though, one of the most important things I learned was how to structure my life to get work done.

Life as an academic seems idyllic—and in many ways it is. My job gives me lots of flexibility to focus on areas of study that interest me. As long as I teach effectively, serve the university, and make a contribution to my field of study, nobody micromanages the research and writing that I do.

But this flexibility comes at a cost. I have to generate my own work schedule and my own deadlines to get things done. There is no boss telling me when a particular study needs to be completed or a specific paper needs to be written.

What I learned from my mentors was the importance of having a regular schedule to get work done. Both of my primary mentors, Dedre Gentner and Doug Medin, were fiercely protective of their writing time. They would carve out blocks of time each day to work on analyzing data and writing up studies. Both of them tried to avoid scheduling any meetings before noon to give themselves the opportunity to concentrate and dive deeply into their work.

The message from their example was loud and clear: It is crucial to make daily progress on long-term goals. A contribution is not made in one sprint. Instead, contributions reflect the accumulation of individual days and particular actions that come together.

This recipe for success is common across many fields. Author

Stephen King has written more than fifty books and hundreds of short stories. That means that he has churned out more than one book a year for his entire adult life. When talking about his writing process, King emphasizes the role of routine. Published interviews with him suggest that he starts working early each morning and writes for several hours a day. He tries to engage in the same set of actions and to treat the process of preparing to write in the same way he treats the process of going to sleep.

As I was thinking about routines, I stumbled on a website called *Daily Routines* that collects quotes from a variety of artists, writers, scientists, mathematicians, and musicians talking about their day, many of whom share similar stories. People who make a contribution take their routine seriously. Many get to work soon after waking up and try to take advantage of the energy they have at the start of the day. Few people really feel like they get a lot accomplished in the middle of the day. Some return to their work late at night. Others use their time later in the day to reflect on what they have done or to spend time with family, or to pursue other interests and hobbies.

What is clear from all of these examples, though, is that successful people have tamed their Go Systems. After all, everyone has temptations in their lives that threaten to derail an overall goal. There are always obstacles to success. The routines that successful people create become regular habits.

Just knowing that people who make a contribution have routines is not enough, of course. You need a recipe for structuring your life so that you can allow your Go System to create more effective habits.

There are three key tools associated with taming the Go System. The first is developing an *implementation intention*. Implementation intentions are specific plans that determine what you are going to do and when you are going to do it.

After you create your implementation intention, you have to grapple with the many obstacles that you'll face on the road to developing new routines. Your Smart Change Journal should already have a list of obstacles in it. As you work to tame the Go System, you will create plans in advance for what to do when obstacles come up.

Finally, you need to ensure that your goals are energized. No matter how good your plan, if you do not get yourself aroused, you will not do anything. As it turns out, though, you have to be careful with how much you energize your goals. It is possible to have too much of a good thing.

How to Get Things Done

Human language is an amazing tool for allowing us to communicate complex thoughts to another person. But language also creates the illusion that you are being more specific about a topic than you really are.

This point was brought home to me for the first time when I was in college taking a seminar on robot planning. I was a cognitive science major. Cognitive science recognizes that the mind and brain are enormous scientific mysteries and that no individual discipline has the single right way to understand how they work. Scientific problems of the mind have to be examined from many different directions. As a result, I took courses in psychology, linguistics, neuroscience, philosophy, and computer science. In addition to taking basic classes on computer programming, I also studied artificial intelligence. One seminar focused on how to get a robot to plan a series of actions.

Imagine you wanted a robot to get from the Empire State Building to the George Washington Bridge in New York City.

At first, you might think about this problem as if you were giving directions to a robot standing in front of the Empire State Building (on Fifth Avenue between 33rd and 34th Streets) to the George Washington Bridge (on the west side of Manhattan at 178th Street). If you go to a website like Google Maps or MapQuest and ask for these directions, you will get a list that describes the route. That route might tell you to drive south on Fifth Avenue to 29th Street, make a right on 29th Street and follow it to the West Side Highway, and then take the West Side Highway north and take exit 14, which has signs for the George Washington Bridge.

These directions are a great description of a route. It is exactly what you would tell someone who was driving from the Empire State Building to the George Washington Bridge.

But it is not nearly specific enough to get a robot from one place to another. After all, the robot needs to figure out which direction is south. It needs to know how to avoid hitting other cars or pedestrians. It needs to know a lot about the rules of the road. It needs to know how to determine its current location and how to find entrances and exits to freeways.

In short, the language we use to describe how to do something to someone else misses a lot of the actual complexity of doing that thing.

Instead, when you talk to people about how to do something, you assume they have a lot of basic knowledge about how to accomplish things in the world. When you give directions, you assume that people know how to drive, walk, or take a taxi. You do not include specific instructions for navigating the world in your description of the route. Indeed, you do not have a particularly good vocabulary for talking about exactly how to carry out specific actions. That's why you do a bad job at describing processes like how to tie a shoelace. It is easier to show someone how to do it than to describe to him how it is done using language.

You do make some changes in the way you talk about things that take into account people's knowledge. Research on the way people communicate shows that they give more detailed descriptions when they think that people have little expertise or familiarity with a situation than when they believe that people know a lot about it. If you are walking down the street in your hometown and someone from out of town stops you and asks for directions, you'll provide a lot of detail. You will mention landmarks along the way and try to estimate the distances people will drive for various aspects of the route. You'll omit most of those details when you are giving the same directions to someone who comes from the same town and knows the area fairly well.

The reason for this long preamble is that when you are trying to change your behavior, you are much closer to the robot that has no idea how to navigate the world effectively than you are to a native of a particular town who is highly familiar with a region. That means that the plans you generate when embarking on behavior change have to be a lot more detailed than the ones you are used to making.

Psychologists Peter Gollwitzer, Gabriele Oettingen, and their colleagues have explored the kinds of plans that people need to formulate. Gollwitzer's work shows that people are most effective when they create *implementation intentions* to prepare them to reach their goals. Oettingen's studies examine the role of envisioning the future using fantasies to help generate those plans.

An implementation intention is a specific plan describing the steps needed to make a contribution. The key to an implementation intention is to create an association between a particular circumstance and the action that is going to be carried out.

If you develop an implementation intention for your trip from the Empire State Building to the George Washington Bridge, you start by identifying a particular location. The Empire State Building is on the west side of Fifth Avenue, so you begin by facing the Fifth Avenue entrance and then turning to your left to ensure that you are moving south. Then you form another intention that when you see the sign for 29th Street, you turn right and continue moving. For each key intersection, you describe the specific situation and the actions you want to take in that situation. You might also develop alternate plans for other modes of transportation like the bus or subway. In addition, you should consider obstacles like traffic that might make it hard to reach your goal and to plan for those possibilities as well.

It seems strange, perhaps, to develop such a specific plan for driving a route, but this plan becomes more important when you are trying to accomplish a new goal.

When you start to focus on a new contribution, the set of steps you need to take are not obvious. Perhaps the best way to get your journey started is with a fantasy.

In common discussion, the word *fantasy* is used to mean unattainable outcomes that we dream about achieving. People fantasize about playing professional sports, winning the lottery, or dating an attractive movie star.

In fantasies like these, people focus on a series of events in which the world is different from the way it is now. Our imagination allows us to think about the world as it could be and not just the world as it is. And in cases like these, the fantasy enables people to enjoy the outcomes that would come from this world.

This same mechanism can be put to powerful use in creating implementation intentions and energizing those intentions to get started making a contribution.

Go back to your Smart Change Journal. At the end of the previous chapter, you created a list of actions that form achievements that will lead to your contribution. Now is the time to take your list of actions and to turn it into an implementation intention. And that requires a fantasy.

Think about your life as it is right now. Imagine how you would fit the actions you need to take into your life. Determine specific times and places where that action would happen as well as the resources you will need to carry out the actions. Imagine what your home or office would look like as you take this action.

When I wanted to learn to play the saxophone, I knew I would need to practice daily. But, making that happen required a more specific implementation intention. On weekdays, I cannot practice until after work. But after work each day, dinner has to be prepared and eaten and dishes need to be cleaned. On many days, my kids needed help with their homework. My implementation intention was to practice after the dinner dishes had been done and the kids' homework was under control. So I planned to start practicing sometime between 8:30 and 9:00 p.m. on weekdays.

I also had to think about where I would practice. The saxophone is a loud instrument, and it requires some space to play it. It also requires some privacy. If people are constantly walking by the practice space, it is hard to concentrate. Also, when you learn to play a musical instrument, you make a lot of mistakes (not to mention strange and unmusical noises). You can get self-conscious about your practicing if you are worried about other people's reactions to the sounds. I added to my implementation intention that I would set up my instrument in the garage to be out of the way of other people in the house.

Notice what happens when you actually start to create a fantasy. Your knowledge about reality starts to constrain the plan. As you think about where to fit the action into your actual life, you begin to make decisions about how the action could actually get performed.

Such a fantasy has two benefits.

First, you begin to discover some of the unexpected obstacles to carrying out your action. When planning to play the saxophone, I had to figure out a time of day that would make sense for regular practice. The aspects of my life that could not be moved (like work, dinner, and child care) had to be put onto the schedule first. Only then could I find the open spaces in the schedule to fit my practice. I had to think about the location for playing and to envision all of the factors that might make each location a good or bad one for playing the sax.

Second, the fantasy forces you to be very specific about the action itself. In the list of actions you created in the previous chapter, it was fine if many of them were stated abstractly. It was enough for me to say "practice every day" to begin thinking about my goals. In reality, an implementation intention needs to be a lot more specific than that. It has to contain information about precisely when and where the action will be taken.

So after the fantasy, the plan goes from "practice every day" to "set up the saxophone in the garage between 8:30 and 9:00 p.m. on weekdays and play for at least thirty minutes. On weekends, practice some time during the afternoon." Of course the implementation intention does not necessarily require an exact time on the clock. The statement of time could be described in terms of situations that come up during the day. For example, you could say "after dinner and before getting ready for bed."

Because you've indulged in a fantasy, you have a better sense that your plan is reasonable. You have already seen ways that the plan might fail, and you have incorporated the potential barriers into your implementation intention. You've thought about aspects of the plan that would not be obvious without first fantasizing.

A specific plan is a great reminder of when to carry out an action. If the plan was just to practice every day, how would you know that it was time to get started? Every time you look at your watch, it is a specific time of day. It would be easy to go a whole day without thinking it's time to practice and only notice the next day that you had not played at all.

So the specific time and context you create as part of your implementation intention remind you when it's time to do those actions. This is particularly important when the action is not one that needs to be carried out every day.

When it became common to install home smoke alarms, most of them were battery operated. Unfortunately, batteries wear out eventually. Smoke alarms come equipped with a mechanism that cause the alarm to sound briefly when the battery is dying as an alert to put in a new one. If you miss the warning, though, the battery will die and the smoke alarm becomes nonfunctioning.

The batteries in a smoke alarm will last between nine and twelve months, so to minimize the chances that they'll wear out, it's best to change them twice a year. The problem is that it is difficult for people to remember arbitrary dates that are approximately six months apart. Fire-safety experts recognized this problem and created an implementation intention for people. They suggested that people change their smoke alarm batteries every time they change their clocks to or from daylight savings time. The nice thing about this recommendation is that it provides a specific time and context to perform an action that happens at long intervals. Because there are always at least a few clocks in your house that need to be changed manually, adding another small job to this list is an easy way to make sure you change your batteries.

Now, it is time to go back to your Smart Change Journal. Your fantasy has helped you develop a set of actions that you perform to achieve your broad goal to make your contribution.

From here, you have to create an implementation intention. For each action in your list, answer these questions: What action am I going to take? When am I going to do it? Where will this take place? How often am I going to perform this action? What aspects of my life will I need to work around to reach my goal? Whose help do I need? What resources do I require?

how can i fit new actions into my life?

What action am I going to take?
When am I going to do it?
Where will this take place?
How often am I going to need to perform this action?
What aspects of my life will I need to work around to reach my goal?
Whose help do I need?
What resources do I require?

After you answer these questions, look over what you have written and see whether you have described the situation specifically enough. Will you be reminded to carry out these steps when the time comes? Are you sure you know how to carry out these steps?

Finally, before moving on, ask yourself whether the actions you are planning are realistic.

One of the first pieces of advice that I give to graduate students in my lab is that the best PhD dissertation is a *completed* dissertation. What I mean by that is that many people have a belief that a dissertation should be an amazing piece of scholarship. This desire to create a true masterpiece can be paralyzing for students. They worry that no idea they have will live up to their standards of excellence (not to mention those of the faculty members on their dissertation committees).

As a result of their concerns, students fall prey to the adage "The perfect is the enemy of the good." As I tell students, if their dissertation is the best piece of work they ever do, that's a shame because they have a long career ahead of them, and it would be all downhill from there. And the members of the faculty who mentor them along the way know that a research contribution is made over time and not on the basis of a single amazing project.

We often set unrealistic expectations for the achievements we make on the road toward making a contribution. True contributions are the result of many small actions, each of which may have some flaw. Only by sticking with the process despite the flaws will you make your contribution.

Unrealistic Expectations for Your Actions

The outcome you desire may not be achievable in the time frame you laid out initially. When I decided to learn to play the saxophone, my hope was that after ten years I would be playing in a band. If I had set the goal of playing in a band within a year, I would probably have been disappointed, because it simply takes longer than that to learn to play the instrument well. You may also need to revise your time estimates after you get started.

A novice runner may quickly go from running for a few minutes to running several miles. After that, she may believe that within six months she'll be able to run a marathon. Running long distances requires a lot of training, though, and after several weeks, she is likely to discover that her time frame was not realistic and needs to be revised.

Careers also move more slowly than you might hope. When you first take a job, you might want to advance quickly to a position of high responsibility. It takes time to move forward. And it turns out that you learn a lot in that time period. Comic strips like *Dilbert* may portray managers as inept functionaries, but most people in management positions have a lot of subtle skills that help them keep their groups performing effectively. Early in a career it can be difficult to see the importance of these skills for handling people. It takes several years of experience to recognize the value of both the technical skills in a job as well as the people management skills necessary to lead a group.

The location for your goals can also be a problem. Think about the where aspect of your implementation intention. Does it make sense in your life? Over the years, I have seen friends who have been seduced into joining beautiful gyms with gleaming new equipment. Often, those gyms are located about a twenty-minute drive from their home or office. It seems like a brilliant arrangement, but quickly, they realize that adding an extra forty minutes into the time for each workout just for the commute to the gym is not sustainable. It would be far better for them to go to a less spectacular place to work out if it will take less time to get there.

Before moving on to refine your plan, take a look at the actions you have described in your implementation intention. Are these actions realistic? Do they seem like things you will actually be able to accomplish? Sometimes, it is not clear whether actions are realistic until after you get started trying to change your behavior, but it is certainly worth thinking about this issue before you commit to a particular way of changing your behavior.

stress and change

If you look at lists of the most stressful life events, one thing leaps out at you. Almost all of them have to do with major life changes. Even positive life changes like getting married or getting a big raise at work are causes of stress.

Change disrupts all of your habits. Once you make a change in your lifestyle, you have to think about all kinds of things that you didn't have to think about before. This extra effort is unpleasant. Also you are wired to be mistrustful of new situations. Familiar situations are probably not dangerous because you have been through them before and you survived. But in new situations, you're predisposed to be extra careful because there may be hidden dangers. When your circumstances change, you are a little more vigilant than you are when you're in your comfort zone.

Stress can get in the way of successful change. As I explore in Chapter 5, stress impairs the ability of the Stop System to do its job. One way to minimize the stress of behavior change is to keep yourself focused. You may be tempted to make lots of changes in your life all at once. If you are going to disrupt your habits, you may as well disrupt them all and make lots of big changes.

However, the evidence is fairly clear that making lots of changes at the same time is a recipe for failure. Staying on top of all of the elements of an implementation intention is hard enough when you are dealing with just one major change. When you try to make several changes at the same time, though, it is difficult to keep track of everything you are supposed to do. As a result, you are much less likely to succeed.

If you feel like there are many changes you need to make in your life, establish priorities. Which changes are the most important? Focus on making the most important changes first. Add new changes to your life only after you have established some new habits and the stress of those initial changes has faded.

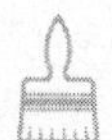

Dealing with Obstacles

To handle the obstacles that threaten to stymie your attempts at Smart Change, you need to envision the obstacles and then plan for them by improving your implementation intention.

You already spent some time envisioning obstacles in Chapter 3. You should have a pretty good preliminary list of obstacles in your Smart Change Journal. Now that you have started to develop your implementation intention, though, you may find even more obstacles than you thought of before. Take some time to add to the list you created. Are there resources that are not readily available to you? Are there people who will get in your way? Are there situations you may encounter that will hamper your pursuit of your goals?

This process of finding barriers to success is a *negative* fantasy. Earlier, I asked you to think about all of the positive things you need to do to succeed. Now you need to take each aspect of your plan and find the reasons it could fail.

After you develop the list of obstacles, you must make this list a little less overwhelming. Start by classifying them into one of three types. Check off each of the obstacles with a different color pen:

Highly likely to happen. These are the obstacles you are virtually guaranteed to encounter.

Unlikely to ever happen. If you are convinced that a particular barrier is unlikely to happen, then it is a good candidate to leave aside for now.

Possibly likely to happen. Chances are there are a few items that fall in the middle; they might happen, but they are not *highly* likely to happen.

Start with the obstacles you think are most likely to happen. Figure out how you can address each one directly. If you think about the situations in which you'll encounter obstacles, you'll be able to recognize them as soon as they occur and to engage a plan that you have created in advance.

Suppose that you are trying to lose weight. You know that you need to ensure that you eat a reasonable amount at meals without overeating. One obstacle to this plan is that you occasionally find yourself at a party or professional function where there is a buffet.

It's important to plan in advance for a situation like this. If you wait until the first time you are at a buffet after you start your weight loss plan, then many things can go wrong. You may be so engaged in a conversation that you use your pre-diet routine for taking food from the buffet. The food at the buffet may be so enticing that it's difficult to hold back from taking more than you should.

If you have a plan already, though, it is easier to engage that plan when you recognize the obstacle coming. You might decide, for example, to start by going to the buffet after everyone has taken a first trip. This way, by the time you finish eating, you will not be tempted to go back for seconds. Or you may want to look for the smallest plate available at the buffet and take only enough food to fit on the central part of the plate without touching the edges. Because you generally eat all of the food you put on your plate, taking a small plate helps control the portion size. Finally, you could decide to eat your meal sitting far away from the buffet to further minimize the temptation to go back to get more.

Make a similar plan for all of the obstacles you think you'll encounter frequently and also look at the list of obstacles that you think may crop up less frequently. It may be worth planning ahead for some of those as well. You don't want to get lost in the planning process, but you do want to be prepared for things that may come up that could derail your chance to make a contribution.

The key is to remember you can always plan ahead when you anticipate a future obstacle.

Keeping a focus on obstacles can be important in a work environment. An entrepreneur of my acquaintance described his strategy for going to networking events. He realized that when he went to these meetings, he would gravitate toward the people he already knew and engage in small talk. These interactions were pleasant, but they were not helping him meet people who might be able to help his business grow.

He engaged in two strategies to support his goal to meet new people at networking events. The first strategy was a general plan. He walked into each event with ten business cards in his jacket pocket. He set the goal to hand out those cards to people he had never met before. That way, he could not spend the entire event talking with the usual crowd. Familiar faces were his obstacles.

The second plan was more specific to each event. He was concerned that if he went to networking events and brushed off the people he already knew, then he might offend a valuable part of his social network. So he would look through the list of invitees and find key colleagues. In the week before the event, he would set up lunch or coffee appointments with those people for the week after the event. This way, when he went to the networking event, he could shake hands with his colleagues and tell them that he looked forward to seeing them soon. This forward planning helped him maintain a good relationship with the network of entrepreneurs in town.

Finally, bear in mind that you will not recognize in advance all of the obstacles you might face. I will return to the issue of obstacles throughout the book to provide other strategies for dealing with the situations for which you may not be able to anticipate.

Energizing Your Goals

Chances are, your relationship to deadlines is a three-act play. In act one, the deadline is far off. You are aware of the project. It might even occupy a place on your to-do list, but you don't spend a lot of time working on it. On occasion, you might put in a few hours, but it is easy to get sidetracked by other more pressing concerns. In act two, the deadline starts to loom larger. At that point, you block off some time in your schedule and get to work. This period is blissful. You work consistently, and you make progress. Often, though, projects require more time than you really budgeted for them.

And that is when act three hits—panic. You clear the decks to focus on the project. But you have so much energy, it is hard to concentrate. You get up and pace around. You long for that feeling you had back in act two when you were immersed in the job rather than being consumed by it.

The structure of this play reflects the way goals are energized in the Go System. As I mentioned in Chapter 2, psychologists call the energy in the motivational system *arousal*. Without the motivational energy provided by this arousal, your goals do not influence your actions at all.

There are three broad ways you can give energy to your goals.

First, there are internal needs and thoughts. Signals from your body can arouse your goals. When your blood sugar level drops, that activates your need to eat. When your bladder fills, that creates the motivational energy to the goal to use the bathroom. These goals are energized directly from the signals of need. These physiological connections allow your motivational system to help keep your body functioning properly.

You can also control your arousal through your thoughts. Thinking about an old friend can energize you to send her an email. Focusing on a project you have at work can give arousal to the goal to get started.

If you really want to see the power of your thoughts to arouse your goals, try a little experiment. Next time you find yourself sitting at home alone, go into your bedroom and lie down. After a minute, ask yourself, "I wonder if there is an intruder in the house?" After a minute or so, you will start getting anxious, and you may even get up and check to see if there is anyone there. Even though you initiated the thought yourself, the motivational system takes over from there once the goal of self-protection is aroused.

The second source of arousal is the environment. Each of your goals is associated with a set of conditions in which actions should be taken. Going to a bar energizes the goal to have a drink. That pesky deadline that we talked about earlier is a conceptual part of your work environment. When you enter that environment, the goals related to that action get aroused.

You want to envision your response to obstacles before they happen so your advance plan can get energized when the obstacle occurs. Without that plan in mind, only the temptation gets energy. I will talk a lot more about the role of the environment in driving your actions in Chapter 6.

Finally, other people are an important source of activation of your goals. You are more likely to be motivated when you see people around you pursuing a goal. There is a lot of evidence that goals are contagious, so when you see someone pursuing a goal, you will automatically engage in that same behavior yourself. Not long ago, I was in a hotel carrying a load of papers and books. Halfway down the escalator, everything slipped from my hands. Immediately, a person at the bottom of the escalator started collecting the stray papers and books, and suddenly several other people joined in. They did not consider their actions carefully. Instead, after seeing one person rush to assist, the goal to be helpful was automatically activated by goal contagion.

How does the arousal of a goal influence your performance?

The three-act play that I used as an analogy reflects an observation made by two psychologists Robert Yerkes and John Dodson in 1908 that is called the *Yerkes–Dodson curve*. They postulated that the relationship between the arousal of a goal and the performance on that goal looks like an upside-down *U*. At low levels of arousal, you're just not that interested in the goal, so you don't put in much effort to achieve it. At high levels of arousal, you have so much energy that you have difficulty staying focused on the task at hand (which I called panic). In the middle, there is a sweet spot where you're able to focus on the goal and get a lot accomplished.

Clearly, then, when you want to engage the Go System to get things done, you want to find a way to reach that great middle section of the Yerkes–Dodson curve. To do that, you need to get to know yourself a little better. In particular, you need to know how much arousal you need to get started working in the first place.

Some people are naturally high-arousal folks. They are the kinds of people who start projects early. They are easily excited by new ideas and quickly concerned about whether projects will succeed. High-arousal people do not need a lot of prodding to get started on a project. However, when a situation promotes a goal strongly, it can push them quickly past the sweet spot on the

Yerkes–Dodson curve to the point at which they are no longer functioning effectively.

Other people are low arousal in their natural state. They need to have a small explosive device set off near them to get them to pay attention to any kind of goal. Low-arousal individuals need to get very close to a deadline before they really start to engage in a project.

What kind of a person are you?

If you are naturally high arousal, then the biggest concern you have is to avoid getting overly excited in a way that impairs your work. There are two big things that will help you achieve your goals. First, set up a structure that allows you to get work done in advance so the pressure of a deadline does not build up too strongly. Second, develop some methods to help you keep calm when the environment gets too exciting. Select a quiet place and do some deep breathing when you feel stressed. Try to keep the environment around you calm, even if there is stress associated with the goal you want to achieve.

If you are naturally low arousal, then your major problem is getting started on something in the first place. You need to ramp up the importance of the goal so you'll get started working on it. As I discuss in Chapter 7, you'll benefit from engaging with the people around you to heighten your motivation. You also need to set early deadlines for yourself to ensure you do not leave big projects for the last minute.

Finally, it is useful to get to know the people around you. It is rare that we do anything important individually any more, particularly at work. Once you're in a group, the members of that group will differ in their natural level of arousal. That means that high-arousal people will want to work on things that have not yet engaged the low-arousal people. By the time the low-arousal people have gotten interested, their high-arousal colleagues are about to have a fit.

When you can, try to match yourself up with people who are similar to you in their pattern of arousal. That will help you avoid situations in which you cannot work with others effectively because of a mismatch in the amount of energy you have for the project. If you cannot avoid working with someone who differs strongly in his base level of arousal, then try to design a work schedule that plays to your strengths. Let the high-arousal person work on the project when the deadline is far off. Save the low-arousal work for the low-arousal person, who works best closer to the deadline.

Arousal and Behavior Change

Have you ever been in the studio audience at the taping of a television show? The audience experience begins well before the actual taping of the show itself. The crowd lines up at the sound stage an hour or so before the show is taped. After everyone files into the studio and takes a seat, there is a warm-up person. This individual is part comedian and part showman. He tells a few jokes to loosen up the mood. He may play some loud music and ask people to dance. He also gets people to practice cheering so that the audience can be heard on the audio track of the show.

Why do TV shows need a warm-up act?

The warm-up increases the arousal level of the audience by raising people's engagement with the show and strengthening the goal to be a part of the program. The warm-up also influences the strength of the emotional experience that people have with the program.

To understand the connection between arousal and emotion, we need to think more about the Go System. As I mentioned in Chapter 2, the Go System involves regions deep inside the brain that developed early in our evolutionary history. The brain regions that support more complex functions like language and reasoning are in the outer cortex of the brain. The parts of the brain that serve these more complex functions evolved (quite literally) on top of the older brain systems.

But these newer brain systems do not communicate that well with the older brain. We have little ability to peer into the inner workings of our own motivational system. Instead, the Go System communicates with the rest of our brain through feelings we experience as emotions.

Our feelings have two broad dimensions: *arousal* and *valence*. Arousal reflects the amount of energy floating through the motivational system. When the system is highly aroused, the feelings are intense. When there is little motivational energy around, the feelings are placid.

Valence is whether the feelings are good or bad. We experience positive feelings in many circumstances, including when we are satisfying our goals. We experience negative feelings when we are not succeeding at our goals.

So the warm-up act is injecting arousal into our motivational systems. That energy emerges in the form of stronger feelings. If the show you watch is entertaining, then the extra energy provided in the warm-up gives you a more intense experience, which you are likely to express with your laughter and applause.

This intensity plays two important roles when you are engaged in behavior change. As I discussed in Chapter 2, when you fail to achieve an active goal, you experience a craving. These cravings are the way that the Go System communicates its failure to the rest of the brain. The more arousal you have in the Go System, the stronger the cravings.

These cravings are unpleasant, so you want to make that negative feeling go away. The easiest way to make the cravings go away is to give in and perform the undesired action. Once that goal is satisfied, the cravings subside. Of course, giving in to the craving takes you further away from the contribution you are trying to make, so you need to find another way to deal with this arousal besides just giving in.

The second reason arousal makes behavior change difficult is that the more strongly the Go System is energized toward a goal, the more the system will direct your mind toward information that will help you to satisfy that goal. As I mentioned in Chapter 2, when you get focused on a goal, the Go System makes objects that will help you achieve your goal more desirable. The more arousal in the Go System, the more powerful these influences.

At the early stages of behavior change, this arousal is most likely to be directed at the goals that reflect the behaviors you want to change rather than the new behaviors you hope to perform. High levels of arousal for an undesired behavior can make it difficult for you to abstain from engaging in that behavior.

To engage in Smart Change, then, you need to deal with the arousal of undesired goals and you need to energize new behaviors. Strategies for grappling with the arousal of the behaviors you don't want to perform will occupy much of the rest of this book.

Chapter 5 explores the Stop System and how to maximize its

ability to help you prevent an undesirable behavior that has been energized by the Go System. Chapters 6 and 7 examine ways to use the environment and the people around you to calm the Go System when an undesired goal has been aroused.

The ultimate goal of Smart Change, though, is to redirect the Go System so the new behavior becomes your new habit. Before the new behavior is a habit, it takes more effort to energize that goal.

To help you perform a new action, build some scaffolding.

If you have ever walked the streets of a large city, you are familiar with scaffolding. When a building is being constructed or renovated, the crew erects a skeleton around the outside of the structure. This temporary work space allows the builders to add to the structure. Once the building has been completed, the scaffolding is removed.

Likewise, new behaviors require scaffolding while you're in the process of creating a habit. Part of that scaffolding comes from the specific statement of your goals that you entered in your Smart Change Journal earlier in this chapter. These specific conditions for performing actions allow a goal to be energized when the right circumstances arise.

Your environment and the people around you are great sources of scaffolding for new behaviors. In Chapter 6, I examine ways to change your environment to energize your goals and to create new consistent mappings that allow habits to emerge. In Chapter 7, I discuss ways to use the people around you to keep you focused on the new behaviors you want to incorporate into your life so you can to resist temptations.

Finally, Chapter 8 helps you make the transition from thinking about achieving your goals to actually working toward making a contribution. As you shift from thinking to doing, you need to energize your goals.

For now, the key is to remember the new behavior will not grow without your creating some kind of structure to support it until the Go System learns a new habit. The purpose of this structure is to energize your goals.

The Takeaways

The crux of behavior change is that your Go System currently directs you toward an undesired behavior. The undesired behavior gets energized by the environment, and that arousal triggers your old habits.

To allow the new behavior to take shape, you have to start by creating an implementation intention—that is, you have to establish a plan for getting things done. A good implementation intention makes specific reference to circumstances that will arise when the behavior should be performed. Specific implementation intentions allow the world to energize your Go System.

Just generating a plan is not enough, though. Inevitably, there will be obstacles in your path to changing your behavior. You have to meet those obstacles head-on by planning for them. Although you cannot envision everything that may get in the way of changing your behavior, you have to generate specific ideas for how you will handle the most common problems you'll face.

Finally, it is important to understand how goals are energized. The Go System influences behavior when goals are given energy. At low levels of arousal, the goal has little influence on behavior.

At moderate levels of arousal, the goal functions most effectively. At high levels of arousal, the system can be pushed into a panic mode, in which behaviors are no longer efficient. It is important to try to keep your Go System functioning in its ideal range.

Once the Go System is engaged, it works efficiently to satisfy the goal that has been engaged. When the undesired behavior is energized, you have to find ways to calm the system so you do not give in to the temptation of the old behavior. You can then use your implementation intention as a scaffold to energize the new behavior until you have repeated it often enough for a habit to emerge.

FIVE

harness the stop system

Finding Your Limits

Creating Distance to Minimize Temptation

Changing Your Beliefs About Willpower

AS LONG AS HUMANS HAVE BEEN WRITING, THEY have been writing about temptation and the difficulty of overcoming it. In Homer's *Odyssey*, Odysseus was curious about the song of the Sirens, those beautiful ethereal creatures who tried to lure sailors to their island by singing. But, alas, the poor sailors who got too close to the island would find themselves shipwrecked on the rocks that surrounded the island.

Nobody was able to resist the song of the Sirens, but Odysseus desperately wanted to hear the song anyhow. So he ordered the crew of his ship to plug their ears with wax. He asked to be tied to the mast of the ship and to be kept there no matter how much he begged to be let go. As the ship approached the island of the Sirens, Odysseus pleaded with his crew to be set free, but they listened to his orders and kept him tied up despite his desire to give in to their song.

This story is a wonderful allegory for temptation. At a distance, you find a temptation tantalizing, and you believe that you are strong enough to overcome it. As the temptation gets closer, though, it becomes increasingly hard to resist until, eventually, you want to give in to it. If you manage to resist (by your own efforts or those of the people around you), then the temptation subsides as you get farther away from it.

Religions also have a lot to say about temptation. The Talmud (a compendium of Jewish thought incorporating the writings of rabbis from the Roman era as well as later scholarly commentary), for example, makes a distinction between the Good Urge (*yetzer ha-tov,* in Hebrew) and the Evil Urge (*yetzer ha-ra'*). The Evil Urge is really a desire to do pleasurable things and to take the easy way out of situations. Some writings talk about the Evil Urge as the urge to engage in the kinds of behaviors that humans share with every animal and the Good Urge as the one that drives people to do things that are uniquely human.

Rabbinical discussion acknowledges that the Evil Urge is extremely powerful. People are engaged in a constant struggle between the inclinations to give in to temptation and to resist it. Some rabbis suggest that learning to overcome temptation takes time. They note that children have only the Evil Urge and that the Good Urge does not really develop until the teenage years.

There is a lot of psychological truth underlying this age-old wisdom. As the Go System gets more energized about a goal, it works harder to help you to achieve that goal. If you are trying to change your behavior, then there will be times when the goal that gets energized is one you don't want to achieve. At that moment, you're in the thrall of the song of the Sirens. The Evil Urge is on your shoulder prodding you to engage in pleasurable actions you have performed in the past but that you would prefer to avoid now.

That is where the Stop System comes in. As I described in Chapter 2, the Stop System is a set of brain mechanisms primarily in the frontal lobes of your brain that act to prevent you from carrying out a behavior that has engaged the Go System.

Unfortunately, the Stop System is not nearly as efficient as the Go System. There are several factors that can interfere with the Stop System as it tries to do its job. Many of them are enshrined in our cultural heritage.

In the classic comedy film *Airplane!*, Lloyd Bridges plays a former pilot. He is called into the control tower at the airport to help someone land a plane in which the crew has gotten sick. A running gag throughout the movie is that the pressure of the situation has foiled Bridges's efforts to kick some bad habits. He starts by saying, "I picked the wrong week to stop smoking!" By the end of the movie, it has also been the wrong week to stop drinking, taking amphetamines, and sniffing glue. This is just an extreme version of the observation that people give into temptation when they are stressed.

Most bars have a happy hour in the period right after work. The idea is that after a whole day of staying under control and buckling down under the watchful eye of the boss, it is time to let loose and give in to temptation. On a particularly difficult day at work, people may drink more than usual, because constant self-control makes it hard to resist new temptations. This is an example of the *ego depletion* effects I mentioned in Chapter 2. A prolonged period of stopping the Go System from doing what it wants to do eventually causes the Stop System to work less effectively. At that point, undesired behaviors can take over.

All this suggests that the Stop System is generally unreliable.

On the other hand, if the Go System gets revved up to do something you would rather avoid doing, all hope is not lost. There are several things you can do to give yourself a fighting chance to avoid giving in to temptation. You just have to learn to harness the Stop System.

Have a Plan

Chances are, you know how dangerous it is to walk into a grocery store when you're hungry. As you walk down the aisles, products seem to leap into your cart. You were determined to buy some vegetables for a salad, a loaf of bread, and some cereal. By the time you get to the checkout counter, though, you also have a pint of ice cream, some potato chips, crackers, and onion dip.

The intuition that being hungry makes you buy more food than you would have otherwise has some data to back it up. Tim Wilson and Dan Gilbert report studies in which they engaged with random grocery shoppers. Before they went into the store, some shoppers who admitted they were hungry were left alone. Others were given a muffin to eat before they started shopping. Not unexpectedly, the hungry shoppers bought more items that they didn't expect to buy when entering the store than those who were not hungry.

The food is jumping into your cart because your hunger is a craving that reflects your goal to eat. The active goal to eat makes the food look more attractive, and your shopping habits often lead you to grab packages of your favorite foods and throw them in your cart without your even thinking about it.

So how can you resist?

The obvious answer (though often an unhelpful one) is to make sure you don't go to the store hungry. And certainly if you have the time to eat something before you head out for your weekly shopping trip, that's great. But what can you do when you have to go into the store hungry?

Let's look more closely at the study by Wilson and Gilbert. Some of the shoppers in their study were allowed to go into the store with a shopping list. Others were not. The hungry people who had a shopping list were much less likely to buy things they didn't intend to buy than those without the list. The shopping list protected people from buying too much just because they were hungry.

Why does the list work?

When you walk into the store without a list, you have to rely on your memory to help you shop. You have a few ideas of what you want to buy, and (you hope) you manage to remember all of those items. At the same time, though, anything else that springs to mind is fair game for purchase. Because you don't have a clear plan about what you are going to buy, anything that looks really good to you stands a good chance of ending up in the cart.

The list is a plan. You do all of your thinking about how you are going to shop before you get to the store. Once you get to the store, you can focus on executing that plan. The reason this works effectively is that it provides a positive action you can take after you stop yourself from doing something you do not want to do.

Generally speaking, in the face of a temptation, you need a plan. The aim of that plan is to lighten the burden on your Stop System. The plan reminds you to avoid the temptation, but then it provides a way of redirecting the Go System so you are not just riding the brakes until they fail.

Return to your Smart Change Journal. Last time you worked on the journal, you thought about the most prominent obstacles you might face. Some of those obstacles are limitations in your resources. You need to think about how to deal with a lack of time, money, or energy that might prevent you from achieving a goal.

Some obstacles, though, come in the form of temptations. Look through the obstacles you have written down already. How many of them are temptations? Now that you're thinking about temptations more specifically, are there other ways that your Go System may be engaged that will stand in the way of changing your behavior? Identify all of the temptations you can.

Where will you encounter these temptations? For the ultimate goal you want to achieve (your contribution), what is the equivalent of the supermarket? What is the environment in which you run into your temptation?

For each of these situations, you need a plan. Like the shopping list, the main function of your plan is to minimize the work that the Stop System has to do by creating a new goal for the Go System in the face of a temptation. When you create your plan, think specifically about the goal you want to engage to dampen the strength of the temptation.

You may also feel as if you were stuck. You know that there is a temptation out there, but you are not at all sure how to overcome it. I suggest a few general ways to help your Stop System in the rest of this chapter. In addition, in Chapter 7, I explore ways to use the people around you to help you overcome temptation. As I discuss in that chapter, you may need to find a mentor who has more experience than you in overcoming the temptations you face. This mentor will help you refine your plan.

Get Some Distance

National Highway Traffic Safety Administration (NHTSA) statistics show that 30 percent of fatal crashes involve alcohol-impaired driving. Between midnight and 3:00 a.m., a whopping 65 percent of fatal crashes involve alcohol.

Despite all of the best efforts to educate the public by schools, the government, and advocacy groups like Mothers Against Drunk Driving (MADD), nearly ten thousand people die as a result of drunk-driving accidents in the United States each year. And these accidents are not a result of a lack of awareness of the problem. According to a 2008 survey by NHTSA, 97 percent of people say they believe that drunk driving is a threat to their personal safety and that of their family.

If people realize that drunk driving is dangerous, then why are they doing it?

If you look at the surveys that are done about drunk driving, just about all of them ask people for their attitudes outside of situations in which they are normally drinking alcohol. Because the respondents are not currently in a bar or at a party, they are mentally far away from the situations in which they drink. This mental distance causes people to treat the question abstractly. People who are answering questions about drunk driving while sitting at work in front of a computer at lunchtime are going to focus on abstract things like the well-known facts that drinking and driving is a dangerous combination and that it is silly to risk your life to get in a car after drinking.

But when those same people are standing in the parking lot of a bar at 10:30 p.m. after having had several drinks, they are psychologically close to the situation. In that moment, they are thinking—assuming they are thinking with any clarity at all—about how they have to get home to get some sleep because they have to go to work the next day. They are worried about the amount of time it will take for a cab to arrive and the cost of paying the cab driver. They are concerned about the logistics of picking up their car at the bar the next day. In the moment, they are mired in the details of the situation. All of the little problems that will be caused by taking a cab home feel overwhelming. And all of those problems can be solved by getting in the car and driving home.

Other factors also kick in when contemplating driving drunk. People are poor judges of how impaired they are when they have been drinking and thus overestimate their ability to drive effectively. In addition, the Stop System is impaired by alcohol. So the goal to get home has an easy time overcoming the Stop System's efforts to prevent you from doing the wrong thing.

In the end, driving drunk is exactly like the Siren song to Odysseus. At a distance, the temptation is not so great. As you get closer in time and space to the temptation, though, it engages the Go System ever more strongly in ways that threaten to overcome the Stop System.

If you want to help the Stop System, create some distance. The farther away you can get psychologically from the temptation, the easier it becomes to overcome it.

There are several ways to get some psychological distance from a temptation.

First, you can create real distance. In Chapter 6, I am going to focus on ways to use your environment to change your behavior. One simple thing you can do is to stay away from temptation. If you want to go out with friends for a few drinks, and you don't want to drink and drive, then leave your car at home and take a cab to the bar or get a ride with one of your friends. That way, when you leave the bar, you simply do not have the option to drive drunk.

Second, you can create social distance. Sometimes when faced with a temptation, it is easier to avoid that temptation when you focus on what a friend might do in the same situation. If you find yourself standing in a parking lot thinking about driving drunk, then ask yourself what you would recommend to a friend who was in the same situation. Because giving advice to someone else creates distance, you are much more likely to give precedence to the dangers of drunk driving than to the inconveniences. If you would tell your friend to take a cab, then follow your own advice.

Third, create some distance in time. One way to use distance in time in your favor is to put together a plan before the temptation arises. Deciding to leave your car at home when headed to the bar not only is a way to use physical space to create distance but also compels you to make that plan well before there is a temptation to drink and drive. So distance in time plays a role as well.

In the face of temptation, you can use distance in time another way. Think about how you would feel in a day, a week, or a month if you gave into the temptation. Imagine that you drive home drunk and get in an accident. How would you feel looking back on it? Allowing yourself to look back on that moment from the future can help you to think about the regret you may experience by giving in to temptation.

Creating distance helps the Stop System in two ways. First,

by thinking of yourself as far away from the temptation, you're helping calm the Go System so that the Stop System has an easier time overcoming it. Second, by focusing on the regret you may experience by giving in to temptation, you are also strengthening the Stop System itself.

When You Lose a Battle, Don't Lose the War

Another staple of comedy movies and sitcoms is the colossal goal failure. There is often a secondary character who is on a diet. This character is smug about her eating choices. She watches everything she eats and makes snide comments about what her friends are eating, and touts her own superior willpower. At some point, though, she eats something she shouldn't, and the next thing you know she can be found shoveling ice cream into her mouth surrounded by candy wrappers and empty containers of take-out Chinese food.

Although this scene is milked for laughs repeatedly, it is quite serious. About 3 percent of women report episodes of binge eating on a regular basis. The cycle of dieting and binging makes it difficult for these women to establish healthful habits for eating.

The basic pattern is straightforward. There is some behavior you are trying to stop. The goal to do that behavior becomes active. So you engage the Stop System and you resist the temptation. But just because you stop yourself successfully, does not mean the goal goes away. In fact, chances are that if you fail to achieve a goal because you have engaged the Stop System, you will experience some cravings as well as some negative feelings. Both of those can get in the way of the operation of the Stop System.

Now here's the problem. The goal has not gone away. So every time you encounter a situation in which you might achieve the undesired goal, that goal will become active again. And because you are engaging the effortful Stop System to try to prevent the behavior, you spend a lot of time consciously thinking about the goal. As a result, it begins to feel as though you were being persecuted by the undesired behavior.

Giving in to temptation can lead to a cascade that has been called the "what-the-hell effect." The Stop System has briefly failed in its role of preventing a behavior. The Go System begins to satisfy a goal that has been held at bay for a while. There is often some immediate stress that comes along with the realization that you have failed at a goal that was important to you.

This initial small failure of self-control turns into a monumental failure. You not only fail, but you fail spectacularly. At times, failing like this can have long-term repercussions creating a return to the old behavior. Dieters may not only break their diet in a specific situation, they may spend several days or weeks eating in ways that undo months of concerted effort.

Ultimately, you want to avoid having a small failure in a specific situation develop into a large-scale problem. In the process of changing your behavior, you will lose a few battles, but you still want to win the war.

To figure out how to avoid the what-the-hell effect, it's important to understand it a little better. And that will require a short journey into values and moral reasoning.

Protected Values

For about ten years, I have been on a plant-based diet. I had several reasons for avoiding meat. The most central was that I wanted to eat differently to feel better and lose weight. Being on a plant-based diet can be awkward in some social situations. Hosts are not sure what to serve you. Not every restaurant has options that are free of animal products. When you attend professional events, the vegetarian alternative to the rubber chicken is often a plate of limp grilled vegetables.

In those situations, when people ask me about my dietary restrictions, I tell them that I am vegan. The word *vegan* was coined by Englishman Donald Watson to help distinguish people who ate a completely plant-based diet from those who avoided meat but would still eat other animal-based products, such as milk, eggs, and cheese. When the term was first created, it was a description of a diet.

Over time, the term *vegan* has taken on ethical connotations. People who call themselves vegan refer not only to their diet but also to their outlook on the relationship between humans and animals. They take a strong moral stand against the use of animals for food, clothing, or research. They are also strong advocates for animals and against many farming practices that give the animals a poor quality of life.

This ethical stance becomes what psychologists call a *protected value*. A protected value is a statement of a rule that you simply will not violate. For a vegan, animal rights and welfare are a protected value. Any practice that can be seen as cruel to animals (including farming, circuses, and medical research) is a violation of that value. For these individuals, there's no prospect of a trade-off. The reason I am uncomfortable calling myself a vegan is that—while I am concerned about animal welfare—it is not a protected value for me. I have a number of colleagues who do important research on animals. I accept that there are research goals important enough to justify using animals in their work. I am willing to make trade-offs, and so animal rights are not a protected value for me.

One characteristic of protected values is that breaking one of those values leads to a powerful emotional reaction. When someone sees another person breaking a protected value (or even talking about breaking a protected value), it leads to shock, outrage, and indignation. When someone who holds a protected value breaks it themselves, their violation leads to guilt and sadness. Indeed, even thinking about breaking a protected value feels like a violation of that value.

Protected values serve an important self-regulation function. When you adopt a protected value, you bring to bear a powerful set of mechanisms that helps you control your behavior. The protected value engages both the Go System and the Stop System. While the Stop System works to prevent you from thinking too much about the issue, the Go System is engaged in directing anger and argument at people who violate the value as well as at the situation that created the opportunity for a violation.

The protected value system helps cultures create rules that are difficult for members of that culture to violate. These kinds of rules are useful for keeping people from doing things that might be in their own short-term interest but would be bad for the group as a whole. That is, we employ protected values culturally as a way to help us deal with the trade-off between short-term temptations and long-term contributions.

If many people in a culture subscribe to the same protected value, then it can help minimize the undesired behavior across the group. For example, the Amish living near Lancaster, Pennsylvania, believe they can serve God through simplicity and community. They hold preservation of the community as a protected value that is enforced by limiting the use of technology. Their regular mode of travel is a horse and buggy rather than a car. In this way, Amish cannot travel far from their community and families. They do not keep phones in their homes (though they may use communal phones for business), so they do not replace face-to-face conversations with the phone.

The community can react angrily to individuals who break these protected values. When members of the community adopt the protected value to avoid technology, they feel guilty if they even consider the use of technology. The group also helps keep temptations out of the community, so people do not face a constant reminder in their local environment of things that might cause them to stray from their values. For all of these reasons, the Amish community continues to thrive in Pennsylvania, despite the rapid pace of technological change experienced by their neighbors.

A potential problem with protected values is that it's hard to have more than one of them. If a single person holds more than one protected value, there is the prospect that these values will ultimately conflict. Many opponents of abortion hold the preservation of human life as a protected value. In their view, no circumstance is important enough to allow a fetus to be terminated. Some abortion opponents also have strong views on the punishment of criminals and believe that murderers should be given the strongest possible punishment. This protected value can come into conflict with the belief that human life must be preserved if possible punishments for murder include the death penalty. These individuals must find some way to resolve a conflict between these two protected values.

This same conflict can play out across individuals. One reason the abortion debate in the United States is so contentious is that people on both sides of the argument often have protected values. One side has a protected value for protecting the lives of unborn children. The other side has a protected value for establishing the right of women to make their own individual reproductive choices. It is hard for these sides to have a conversation because each side engages moral outrage at the thought of breaking their own value. Consequently, people on opposite sides of the debate about abortion have difficulty communicating; the entire discussion requires thinking about violations of protected values.

Avoiding the What-the-Hell Effect

What has all this talk about protected values got to do with the what-the-hell effect?

When people are struggling with a particular temptation, they often create powerful rules to help them deal with the temptation. These rules engage the protected value system. They lead people to avoid thinking about the temptation as much as possible. They also create powerful feelings of guilt when the temptation overcomes the Stop System.

When a community adopts a protected value, the community

works together to help each member avoid temptations, both by using social disapproval and by helping rid the environment of temptation. When an individual person is struggling with a specific temptation, though, the community may not be focused on that person's struggle. And because the community at large is not focused on the temptation, the environment continues to call attention to the temptation.

A woman on a diet, for example, will find tempting foods that do not fit her diet no matter where she travels. She is dieting, but other people around her are not. Consequently, the woman will often be activating the tempting goal. The protected value system will engage strongly to drive her behavior away from the temptation. Because the protected value system creates arousal, it is easy for the dieter to get agitated. If she gives into the temptation, then the energy in the motivational system may strongly energize the goal to eat, which leads to a binge. At the same time, the guilt of violating a protected value causes the dieter to want to avoid thinking about breaking the diet, and so the cycle continues.

To avoid the what-the-hell effect, then, it is crucial to break this protected value cycle. There are several ways this can be done.

Start by recognizing when you are in this cycle. You know you have a protected value when you experience a high level of guilt every time you fail, and you feel agitated or angry when you even start to think much about the temptation. When you have a protected value, you do not even want to think about possible violations of your value. As a result, you're likely to avoid planning for situations in which you might fail because you feel uncomfortable with the topic overall.

You might think that having guilt would be a strong motivator for avoiding a temptation. When you control your actions by creating the conditions for guilt, though, you are setting yourself up for spectacular failures.

The purpose of protected values is to ramp up the strength of the Stop System to help you overcome a temptation. Although this may be effective in the short term, it is important to activate the Go System to learn new behaviors if you are truly going to engage in Smart Change. That means you have to find positive actions you can take to energize your Go System. In Chapter 3, I recommended that you focus on positive goals. The reason for that suggestion was that the Go System can learn only when you carry out an action. A constant use of the Stop System, even when enhanced with a protected value, will not dampen the reaction of the Go System to the temptation.

Another reason for using the Go System is that in the long run the Go System can actually make temptations less troublesome for you. Research suggests that when people have developed good habits, the presence of a temptation actually activates the desired positive goal more strongly. Someone who has successfully changed his eating behavior in a more healthful way will actually activate his goal to eat healthfully when faced with a tray of wonderful desserts.

Finally, you have to remind yourself that behavior change is hard. Failure is not a sign of weakness; it is a part of the process of behavior change. Some days, the Go System is going to win, and you are going to give in to your temptation. If you treat failure as a part of the process of behavior change, then you realize that a series of successes punctuated by the occasional small failure moves you slowly toward your contribution. It is only when you allow a small failure to mushroom into a huge one that you run the risk of putting your contribution in danger.

Protected values are a problem because they draw a line in the sand. In an effort to strengthen the Stop System, they equate the battle with the war. When you are unwilling to accept any trade-off or compromise, any small failure is magnified because it feels as though it had put your contribution in doubt rather than just getting in the way of a particular achievement.

Instead, you have to take the long view. Slow and steady progress does not mean that there will never be any setbacks. Some days, it will feel as though you could not possibly make progress on your contribution. On those days, take a step back from the situation. Create a little mental distance. Remember that you did not create the behaviors you are trying to change overnight, and they are not going to change overnight either.

Call on Your Reserves

The overriding message so far in this chapter is that the Stop System has limited resources and is generally fallible. While that is true, it is also important to bear in mind that your beliefs about the effectiveness of the Stop System will actually affect how well you are able to stop yourself from giving in to temptation.

The research of Carol Dweck and her colleagues focuses on the distinction between *talents* and *skills*. Talents are abilities that you are born with. You may be able to improve on those talents with practice, but your ability to achieve is limited primarily by your level of talent rather than the effort you put in. Skills are abilities that depend mostly on effort. Even if you start out doing these things badly, with enough time, effort, and practice, you can come to be expert in them. Dweck considers whether you think various abilities are talents (in which case you hold what she calls an *entity* mind-set about the ability) or a skill (in which case you hold an *incremental* mind-set).

Research suggests there is a strong skill component to almost anything in life. That is, you can improve almost anything with hard work and practice. This is true for things ranging from music to art, sports, intelligence, and computer programming. For any of those abilities, you may have some natural talent. When you first sit down at a piano, you may immediately be able to pick out a song or it may take a while before anything resembling music comes from it. But your ultimate ability to play that instrument will depend primarily on the amount of time you spend working at it and not on some innate talent.

Even though almost everything is a skill and not a talent, people tend to believe the opposite. When we see a great musician play a concert, we marvel at her talent. When we read about a new invention, we are impressed by how smart the inventor had to be. We assume that these individuals had some spark that we ourselves did not have. And that spark allowed them to achieve greatness. In other words, we often believe in talents when we should believe in skills.

From what I have said so far in this book, it would seem that the Stop System is a great example of a talent. You have some fixed ability to engage the Stop System, and when it is pushed to its limits it is bound to fail. It turns out, though, that your beliefs about the Stop System affect how well it works.

Research suggests that if you adopt an incremental mind-set about the Stop System and treat it like a skill, then you are better able to overcome temptation than if you have an entity mind-set.

That is, if you believe that willpower is a bit like a muscle that can strain to overcome difficult temptations and can get stronger with use, then you do a better job with temptation than if you believe that your capacity for willpower is fixed.

What does this mean? Why have I focused on the limits of the Stop System if you are really able to overcome any temptation if you want to?

It is important to realize that no matter how much work you put in to shore up your Stop System, it has to be a secondary tool in behavior change. The only path to lasting behavior change comes from allowing the Go System to engage a new set of goals that support the new and desired behaviors at the expense of the old and undesired ones. As a result, I have emphasized the importance of the Go System and the fallibility of the Stop System. If you tax the Stop System too much, eventually it will fail.

That said, your Stop System is probably more powerful than you think it is. Studies suggest that just knowing that willpower is not a fixed resource (as an entity mind-set might imply) helps people overcome a temptation. And it is true. Even though willpower is limited in the long run, you do have the capacity to stop yourself from carrying out behaviors that you want to avoid in the short term. When you face a temptation that seems too difficult to overcome, just remind yourself that you are stronger than that temptation.

More important, the tools in Smart Change allow you to overcome a temptation both by engaging the Stop System and by influencing the Go System. In this chapter, I focused on techniques like getting some distance on the situation to dampen the activity of the Go System. In Chapter 6, I examine ways to structure your environment to prevent behaviors you want to avoid and to provide scaffolding for the new behaviors you are trying to encourage. In Chapter 7, I present ways to engage with the people around you to help turn your new behaviors into habits and to supplement your own individual Stop System by offloading some responsibility for preventing behaviors to other people.

The Takeaways

The Stop System is much less efficient than the Go System. It can be disrupted by factors like stress, alcohol, and overuse. While the Stop System can allow you to overcome more temptation than you might think you can, it is not the ideal way to prevent undesired behaviors as you develop new ones.

At times, you may try to strengthen the Stop System by engaging protected values that make a behavior taboo. These protected values can increase the power of the Stop System by making a violation of that value so threatening that you'll want to avoid thinking about the behavior and will experience guilt when thinking about it. However, protected values elevate the failure of a specific accomplishment to the status of the failure of an entire contribution. As a result, engaging protected values can lead to the what-the-hell effect in which people fall off the wagon after experiencing a single goal failure.

To give the Stop System the best possible chance to succeed, start by having a plan. Use your Smart Change Journal to plan for the temptations you are likely to face. What will you do in those circumstances? When you're prepared in advance, it's easier to engage in behaviors to dampen the activation of undesired goals and to engage positive goals that will interfere with the temptation.

Finally, remember that distance influences the strength of a goal. When you are close to a goal, it is far more active than when you are far from it. When you find yourself in the grip of temptation, find ways to increase your distance from the undesired goal. You can change distance in space, time, or social relationship. Each of these kinds of distance can decrease the strength of a temptation.

SIX

manage your environment

Support New Behaviors

Create Habits

Block Temptations

CHANCES ARE, YOU BRUSH YOUR TEETH REGULARLY. This habit seems so deeply ingrained in people's daily routines, it is hard to imagine going through life without regular tooth brushing.

Indeed, some type of tooth care has been done for millennia. There is evidence of tooth brushing in ancient Egypt, in ancient Greece, and in India. By the eighteenth century, brushing teeth had become a regular part of the daily routine for many people in Europe. The United States lagged behind this trend. Although companies started to mass-produce toothbrushes and tooth-cleaning powders in the United States in the late 1800s, it did not become a widespread habit in this country until the 1940s. Soldiers in the U.S. Army during World War II were required to brush their teeth, and they continued that habit when they returned from the war. Improvements in toothpaste such as the addition of fluoride, which greatly reduced the incidence of cavities, also led more people to brush regularly.

How does this kind of widespread habit change happen?

For one answer, take a look inside your bathroom. Chances are, you have designed your bathroom to support your habit to brush your teeth. Many bathrooms have porcelain fixtures built into them that hold a toothbrush and even provide a place for your toothpaste. If you don't have something built in, then you have probably bought some kind of device to keep your toothbrush next to the sink.

Because of this organization in your bathroom, you can engage in your daytime and nighttime routines without having to think much about them. Standing in the bathroom in the morning, you see your toothbrush. The combination of the morning, the feeling of a dirty mouth, and the sight of the brush engages the Go System to start your routine to brush. All the while, you can focus your thoughts on more important things like your plan for the day.

Although tooth brushing has become a widespread habit, regular flossing is less common. Almost everyone owns some dental floss. Dentists often give out free packages of floss to their patients. And people will use floss when something uncomfortable like a sliver of popcorn gets caught between two teeth. But there is much less universal compliance with daily flossing than with daily tooth brushing, even though dentists recommend both.

There are several problems with flossing that make it less likely to become a habit than brushing. It is generally messy. You have to stick your fingers in your mouth, which is less appealing than using a brush, and so people do not repeat it often enough to engage the Go System to create a habit. Also because the primary reward for flossing comes in the long term, if you forget to floss, there is little to immediately remind you. Only when you visit the dentist with puffy gums do you get the powerful signal that you should have flossed.

Perhaps the most important problem with flossing is the floss container itself. Floss is usually sold in rolls that are placed inside a small plastic container with a dispenser and a device for cutting off a length of floss. The containers themselves may be pretty, though they are often bland clinical white packages. There is no standard size or shape for the package. As a result, it's not clear where to put it in your bathroom. Many people put their floss packages either in a drawer or in the medicine cabinet behind the mirror.

Neither of those spots is a visible part of the environment. As a result, you don't have a good visual reminder that it's time to floss. It's hard to develop a habit if there is no information in the environment to promote that habit.

When you try to change your behavior, there is a tendency to focus on your own psychological characteristics. You want to remember the new behavior. You want to use your internal defenses to avoid temptation. You spend time looking inward to find ways to change yourself.

However, successful behavior change also requires looking outward. Your environment is a powerful driver of what you do. Because your habits involve a consistent *mapping* between the environment and a behavior, your habits are activated by the world around you. Do not assume that behavior change is a purely internal structure.

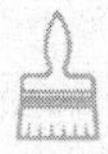

Learn to Manage Your Environment

To use your environment to affect change, you need to structure your world to support the creation of new habits. Also, you need to use your environment to block temptations. And, finally, you need to manage your environment to help you deal with obstacles.

GETTING TO HABITS

In 2004, Procter & Gamble (P&G) launched a new product called Scentstories. The idea behind it was ingenious. The problem with traditional air fresheners is that they have an influence on you for only a few minutes, because you quickly adapt to smells in your environment. If you spray an air freshener in a room, the smell quickly fades from your awareness. That does not mean that the chemicals in the air that cause the smell are gone. You just don't notice them anymore.

To help keep the smells noticeable, P&G created discs roughly the size of a CD that could be put in a device that would "play" them. Over a period of thirty minutes, the disc would release five different smells. Because new smells would appear every few minutes, you would not adapt to one particular smell, and you would continue to experience enjoyable smells in your house for a full half hour. The idea was that discs could actually tell a scent story like taking a hike through a field and forest. The smells would be related to what you might experience on that hike.

The product worked quite well (and I think it was a pretty neat idea), but it was a colossal failure. Few people bought the base unit, which cost about $30, and even the people who bought it rarely used it, so they did not buy the scented discs.

With any real product launch, it is hard to know exactly why it failed. The expense of the base unit was certainly part of the problem. A more important problem, though, is that the product was poorly designed to help people create and maintain a new habit. After all, a product like this can succeed only if it induces people to buy more discs every few months.

The biggest flaw in the design is that the unit itself resembled a CD player. It was a white-and-blue plastic contraption with modern-looking curves, a translucent top, and several buttons and LEDs on the front. When you first see it, the product looks quite sleek, and you can imagine why the people who created it were pleased with its form.

Think for a moment, though, where you keep your CD player. Chances are, it is off to the side of a room or in a stereo cabinet. If you want to listen to music, you walk over and turn on the CD player, but otherwise, it is out of sight. And that is fine for CD players, because you probably have a set of routines for listening to music. You may even keep CDs in a more visible spot that will remind you of things you might want to hear.

You don't have a routine for playing discs of scent, though, because that is a brand-new behavior. If P&G wanted people to learn a new behavior, then it was not a good idea to have a product that people were likely to put in a corner. Instead, the product should have been designed so that it could be placed in a more prominent spot. If it looked like a planter, for example, then it could have been put out on an end table next to a sofa or on top of a credenza.

Putting the product out in the open would allow it to serve as a reminder that it was available for use. The environment serves as a reminder of what we can accomplish. To develop a new behavior, it is crucial to use the environment to help.

USE THE ENVIRONMENT TO SUPPORT A NEW BEHAVIOR

It is important to keep the end in mind. The ideal state for Smart Change is the creation of a habit that will allow the Go System to perform the new behavior automatically. The key is to create a consistent mapping between the environment and a behavior and then to repeat the behavior in that environment.

To begin, make sure there are key reminders of the new behavior in the world. When you are first learning to perform the behavior, you need reminders of when it should be done. Information in your environment that you can see or hear provides the kind of scaffolding discussed in Chapter 4 that enables you to remember to perform the behavior when you are first learning to do it.

If I were improving Scentstories, I would recommend that the product be designed so it could occupy a prominent place in the world. Sometimes, however, the object itself cannot be placed in the environment, so another reminder is more appropriate.

Here's an example of that situation. Over the past several years, there has been a concerted effort to reduce the number of disposable bags used at grocery stores. The stores themselves began to sell reusable bags. Many cities and towns have begun to ban the use of plastic bags or to significantly limit their use.

A key problem with this new system is that people need to remember to bring their reusable bags into the store. In suburban areas, people keep several bags in the trunk of their car. Early on, though, people would walk into the store without the bags, and realize they had forgotten them when they got to the checkout lane. Stores responded by posting signs all over the parking lot reminding people to bring their bags into the store. These signs were a visible reminder to support a new behavior.

Eventually (perhaps even by the time you are reading this), bringing reusable bags to the grocery store will be a routine. Even then, it may be helpful for stores to keep some reminders in the parking lot. Once a behavior has been established, it's triggered by the environment. The bag signs in the parking lot are just one of several aspects of the shopping environment that may support the habit of bringing bags into the store.

More generally, it is important to structure the environment so a new set of behaviors can ultimately become a habit. That means you need to examine your environment and think about how you interact with it and where you want to locate key items to ensure you can turn a repeated behavior into a routine. Look for inconsistent mappings. Are there situations in which an item that is crucial for performing a habit gets moved to different locations? If so, is there a way to make the world around you more consistent?

For example, many companies have moved to an open office plan to save money. In this setup, employees work in rooms with a large number of desks. Nobody is assigned a specific desk, though, so someone may work at several different desks each week. One problem with this system is that key desk accessories like staplers and scissors may be placed in a different location at each desk. Consequently, people in this type of office environment cannot develop habits for where to look for everyday items. Instead, they end up having to disrupt what they are thinking about when they want to staple papers or cut something. This might seem like a small issue, but people who work in this environment report being quite frustrated by the constant changes in their daily routine. It would be more effective for the company to establish a template for each desk so people could use the same work routine regardless of where they were sitting on a given day. Indeed, some companies that have adopted an open office plan have provided caddies for desk accessories to help their employees create and maintain habits.

Finally, try to give yourself enough repetitions for a habit to form. As I discussed in Chapter 2, there is no hard-and-fast rule for creating a habit, but after you have done the behavior in a particular environment on about twenty different occasions, you are likely to be well on your way to creating a habit.

Now, it's time to return to your Smart Change Journal. Are there elements of your environment that you can change to support your new behavior? For each of the steps in your implementation intention, think about what the world around you will look like. What reminders can you put in your world to help you perform the behavior? What aspects of the world do you need to keep constant to ensure that you will ultimately be able to create a habit?

Picture the Key Environments Related to the Changes You Want to Make

One thing that can be helpful for this exercise is to sit in the spaces where you are most commonly engaged in the behavior you are trying to change. Take a picture with your phone or a digital camera. Draw a diagram of the space. Spend some time thinking about how you would like to set up your world to support your new behavior. Tape a photo into your Smart Change Journal or draw your own sketch. The aim is to do some advance planning to prepare your spaces to support the actions you ultimately want to take.

Changing the Environment Disrupts Old Habits

Because habits link the environment to behaviors, changes in the environment can prevent you from performing habitual behaviors. For example, in the spring of 2002, the psychology department at the University of Texas moved into a new building. For years, the department was spread across several buildings on the south mall of the campus within sight of the famous UT tower, which is a familiar symbol of the school. The new facility, on the north end of the campus, was beautiful: offices with wood and brushed metal furniture and new built-in bookcases; spacious labs with high-speed computer networks and state-of-the-art laboratory facilities; elegant stairwells with windows that brought in a lot of natural light.

Yet, for several weeks, everyone in the department was cranky. If you asked, everyone had only good things to say about the new building. Members of the faculty were pleased with their offices and labs. Students raved about their new work spaces. Staff got spacious locations to do their work.

The problem was that everyone had to think about every part of their daily rituals. There was no aspect of their old routines that was preserved. We had to use a new parking garage, so many of us had to change the route we took when driving to campus. We had to think about where the mailroom was rather than just walking there on autopilot. We had to find new paths to walk to other locations on campus. We changed the places where we got coffee, picked up a breakfast taco, or grabbed a sandwich for lunch. We would find ourselves standing at our bookcases searching for a particular journal that we used to be able to grab without thinking.

Everything had suddenly become an effort.

For a group of people who wanted to focus on work, it was unnerving to have to think about these mundane life details. When you want to keep your head in the clouds, it is a pain when you keep being dragged to Earth.

But this same kind of frustration can sometimes be turned to your advantage. When you are trying to make a change in your behavior, a little life disruption is just what you need.

The hardest habits to change are the ones in which your Go System runs on autopilot. There are already strong associations between aspects of your environment and your behavior. Without making any changes, you are going to continue to want to persist in your habits.

Change Your Environment, but Only Just Enough

Rearrange your environment in a radical enough way that you cannot engage your habits directly. Imagine, for example, that you ran a company that was introducing a new product—say a new kind of tomato sauce. Most shoppers already have habits related to the products they buy. They might have one brand of tomato sauce they use on pasta, another that they like for lasagna, and a third that they put on eggplant. When they get to the store, they know what brand they want to buy. They know where to find it on the shelf. They walk to the wall of tomato sauce at the supermarket, reach out, grab the familiar jar from its typical location, pop it into the cart, and move on.

These habits are a problem if you are trying to introduce a new brand. How do you get shoppers to stop for a moment and notice the new kid on the block? No matter how much advertising you do, if shoppers stay on autopilot in the supermarket, you're going to have a hard time breaking through.

What if you could get the supermarket to rearrange their shelves? (It is possible if you're willing to pay a big enough placement allowance, but that's another story.) If the sauces that were usually on a high shelf were now placed toward the floor, and those on the floor were moved higher, then shoppers would no longer be able to use their habits to reach out and grab jars without thinking. By disrupting the environment, you're forcing shoppers to think. And when they think, they have the chance to consider making a different purchase from normal. They have a chance to go against their habit.

You want to do the same thing in your own life to support the development of new habits. When you are truly ready to change your behavior, you want to force yourself to think about your current behavior as much as possible so you're not acting purely based on what you've always done.

Suppose you were trying to change some of your behaviors at work to make yourself more productive. Perhaps you spend too much time checking your email and surfing the web and not enough time reading the latest articles in your field and putting in concerted effort working on long-term projects.

When you are ready to change your behavior, consider getting a new computer with an updated and unfamiliar version of the operating system. You could also change the web browser you use regularly. You might even want to get a different word processor or piece of presentation software.

In the short term, these changes will be frustrating and will make you feel like you were being less productive. However, these changes remove your ability to do everything by habit. Now you can think explicitly about whether you really want to log into your email program first thing in the morning or whether you would rather get some writing done before getting to your correspondence. You can make new decisions about how to configure your information technology environment. Perhaps it's finally time to ditch that instant messaging program that is continually interrupting you from the projects you're struggling to finish. This period of disruption gives you a few uncomfortable weeks in which you can change your habits, eliminating some old behaviors and adding some new ones.

While you are changing your environment, you can also take the opportunity to protect yourself from yourself. In Chapter 1, I described the classic experiments on delay of gratification by Walter Mischel and his colleagues. Those studies are nasty because the researchers call kids' attention to a temptation, and then they leave that temptation in the room so the child must deal with the ever-present temptation, unsupervised for fifteen minutes.

If you have a behavior you are trying to change, you do not have to face your temptation on a regular basis. When you engage in Smart Change, you want to set up your world so the things you don't want to do are hard and the things you want to do are easy.

What Adjustments Can You Make to Your Environment to Help the Changes You Want to Make Become Habits?

In Chapter 1, I introduced you to Mike Roizen, the wellness director for the Cleveland Clinic. Thanks to his efforts, smoking rates among employees dropped significantly. One of the most important policies that the Cleveland Clinic instituted to help curb smoking was to make the entire medical campus smoke-free. Thirty years ago, it was common for people to be able to smoke in their offices and in restaurants, bars, and other public places. Starting about twenty years ago, more public spaces began to institute smoking bans, which helped reduce the number of people who smoke because smokers had to leave the building. This small inconvenience meant that it was harder to smoke purely by habit, because a smoker had to get up and walk outside to have a cigarette. It also limited the number of cigarettes people would smoke because each one would require a break from work or a disruption of a conversation at a restaurant or bar.

Banning smoking from an entire medical campus goes a step further. Now employees may have to walk several blocks to have a cigarette. Each smoke becomes a major ordeal. The harder it is to engage in a behavior, the less attractive it is continue doing it.

The Cleveland Clinic also makes good behaviors easier to perform. They host regular farmers' markets on their campuses to make it simpler for their employees to get fresh fruits and vegetables. They offer free yoga classes for any group of four or more employees who want a class to help them reduce their stress and get some physical activity.

Dealing with Others Who Share Your Environment

When you're trying to change your environment to make it more or less supportive to eliminating or creating habits, you are probably not the only one who has control over elements in your environment. For example, if you want to avoid high-calorie foods you may avoid bringing sugary snacks and drinks into your home. If you don't live alone, though, then you're imposing your own dietary changes on the rest of the household, which is not always easy.

When you have to involve others in changes to the environment, it is important to reach a consensus that everyone can accept. It is crucial to manage your environment, but it is also important to have help from the people around you when you are trying to change your behavior. (The role of other people in helping you change your behavior is the central topic of Chapter 7.)

Consensus is important because resentments can get in the way of your attempts to change your behavior. Bad social interactions cause stress, which can interfere with your ability to avoid temptations. When the people around you resent the changes you have made, they will find ways to undermine your resolve to continue your behavior change. When you know you have taken actions that bother the people you live with, it can sap your motivation to follow through with your plan to change your behavior.

To help achieve a consensus to change the environment, start by talking with the people around you about the importance of your goal to change behavior. Lay out your ideal plan for the environment. Let everyone know your priorities for the changes as well as the other changes that would be helpful, but are less crucial. Then allow them to talk about their objections to the changes. Be willing to compromise to make the changes you think are most important to helping you reach your goal of Smart Change.

A common example involves diets. When one member of a family goes on a diet, it feels like the rest of the family has to go along for the ride. Suppose you decide to lose weight. It can be hard for you to get support from others if you are also influencing what they can eat. Start by identifying your biggest temptations. (For me, it was always ice cream.) Try to remove the temptation from your house. If you cannot keep it out of the house completely, perhaps you can have special snack nights when you bring that temptation home for the rest of the household to enjoy. Maybe you can create an event out of snack night. Do things that recognize that the changes you are making may be good for you but are difficult for everyone else. The work you put in to include everyone in your plans will pay off in higher levels of support for your behavior change efforts.

Remember the most important thing you can do when preparing to make changes is to discuss your plan openly before you try to put it in place. People are often much more willing to do things when they are asked than when they are told to do them. They feel better about changes when they are given the option and agree to them than when the choices are forced on them.

Finally, if you are going to impose on people's environment to support your own changes in behavior, present your suggestions for the changes you want to make and then give everyone a day or two to think about it. Often, people can react negatively to changes when they are first presented because those changes often involve some kind of loss (such as a loss of freedom or ready access to ice cream). Decades of research studies demonstrate that losses can be felt quite strongly. Giving people the time to adjust to a potential loss can make the next set of conversations more constructive.

As you work on your Smart Change Journal, think about the other people you interact with who might have objections to your ideas. What objections might they have? Are there alternative ways of changing your environment that might be better for other people, while still being effective for you? Work through these possibilities before you sit down with others to start negotiating the changes you want to make.

Dealing with New Environments

So far, I have talked about two roles the environment can play in helping you change your behavior: Use the environment to help the Go System achieve your new goals, and manipulate your environment to minimize the burden of old behaviors on the Stop System. The tools I have discussed so far, though, are focused on locations where you have some control. You can rearrange your own kitchen, but your coworkers might not take it too kindly if you rearrange the break room at work just to help yourself develop new eating habits at the office.

To develop strategies to deal with new environments, we need to spend some time exploring the way the body and mind are interconnected, or, in the parlance of cognitive science, *embodied cognition*.

To put this concept in perspective, we have to go a little further back into the history of psychology. In the 1950s, psychology in the United States entered the *cognitive revolution*. Psychologists, linguists, and computer scientists started to forge a new way of thinking about the mind, based on advances in our understanding of computers. The idea was that we could think of the mind as if it were a computer.

It is no surprise that scientists hit on computers as a way of thinking about what the brain is doing. There is a long history of using the most complicated machines of the time to understand the way the brain functions. In the 1800s, when steam engines were the most complex devices invented, there were hydraulic theories of the mind that used principles of heated fluids in containers to try to understand the drives and forces that caused people to act. Because computers were first being developed in the 1940s and 1950s, it made sense to start thinking about how the mind was like a computer.

If you think about the computer on your desktop, it probably seems a bit like a brain in a box. You type to your keyboard, which sends messages to your computer, telling it what to do. The computer responds by changing what you see on the display or perhaps by sending a document to the printer or an email to another computer. That is, it feels as if the computer were taking in messages, working on them, and then sending out other messages to other devices.

You might assume that, like the computer, the brain receives signals from the eyes, nose, ears, and skin and uses that information to calculate what should be done next. From there, messages are sent to the mouth, arms, and legs to take action.

As appealing as it is to think of the brain in this way, it is wrong. The brain and body are intimately connected in many ways. In Chapter 2, we saw how needing to smoke a cigarette could affect your beliefs about how much you want money and cigarettes. There are many ways that the current state of your body affects the way you think. You are much less effective at doing hard mental work when you are tired than when you are well rested. It is hard to think clearly when you have an injury.

The body and mind are connected in more subtle ways as well. Consider your ability to see the world. The amazing thing about sight is that you open your eyes, and immediately you have a rich experience of the world around you. You see objects, colors, and movement. You are able to estimate how far things are from you. All of this experience occurs a fraction of a second after you look at something.

For a long time, researchers exploring the visual system assumed that the primary goal of vision was to help people develop an accurate three-dimensional view of what is happening in the outside world from the images received from the two eyes.

While we certainly do get a feeling for what is out there in the world, vision does a lot more than that. Vision researcher J. J. Gibson pointed out that the main goal of vision is not so much to give an accurate description of the world outside but rather to help you navigate the world. You need to know a lot more about the outside world than just what things look like. You need to know which things you can grasp, which ones you can stand on, which items you can move, which objects you can jump over. Vision has to tell you what you can do with the objects in the world.

To do that, though, your visual system needs to know something about your body. It needs to have a sense of your height and weight. It needs to know how strong you are. Vision needs information about the actions you routinely perform. All of this information about the body is incorporated into your understanding of the outside world.

Your behavior is strongly influenced by all of this information about what is possible for you to achieve in the world. There is a company I visit often that has glass doors separating an atrium from a conference room. When you are outside the conference room, you need to pull the door to open it, but when you are inside, you need to push the door in order to get out. On each side of the glass doors are vertical bars that look easy to grasp. As a result, no matter which side of the door people are on, their first instinct is to grab the bar and pull. Even people who have been working in that office space for years routinely mistakenly pull the door when they are inside the conference room and then have to correct their action and push it.

When you learn more actions and new ways of interacting with the world, it changes the way the world looks to you. One morning, I read a newspaper article about people who were members of a parkour club in Austin, Texas. Parkour trains people to run, jump, climb, and swing on surfaces in order to overcome obstacles quickly. Movies like the Luc Besson film *District B13* and the James Bond movie *Casino Royale* had extended scenes involving parkour in which characters chased each other through urban landscapes.

When someone learns parkour, they exploit small footholds and handholds in walls. They grab onto bars on windows and fire escapes to swing. They look for rocks and other platforms that might allow them to jump to a higher place. A striking thing about the article I was reading was that many of the participants talked about how they saw the world in a different way after learning parkour. Suddenly, they could see places to put their hands and feet or locations for landing from a jump that they never saw before. Their newfound skills changed the way that world looked for them. These people were not speaking metaphorically. Their visual systems provided new information, simply because of the new skills they learned.

It is important to keep in mind that the influence of the body on the mind happens automatically. You do not exert effort to cause your visual system to help you see the actions you can take on the world. It is a natural outcome of the process of learning new actions.

When you find yourself in an unexpected environment, you have to think about how to influence that environment to align the goals you want to achieve with the actions that are easy for your body to do and to make the actions you want to avoid more effortful to achieve.

In Chapter 5, I suggested that if you are on a diet and you find yourself at a buffet, you should select the smallest plate available and use that when you get your food. Why do you want a small plate? When you go to a buffet, you use the plate as a guide for how much food to take. Your bodily interactions with the plate help you fill it, putting on enough food to fill it without putting on so much that the food starts to tumble to the floor. Then, when you sit at the table, you start eating what is on your plate, and you continue until the plate is empty. After you eat, you may choose to get back in line to get more food, but that action requires more effort than just filling up a larger plate with more food in the first place. So you are making the undesired action (taking too much food) harder to perform than the desired action (eating a smaller portion).

There are many ways to influence novel environments in ways that help you to satisfy your goals. In 2001, the U.S. surgeon general suggested that people should try to take ten thousand steps a day to get more physical activity into their lives. (You walk a mile in about two thousand steps, and you burn about a hundred calories for each mile you walk, so walking ten thousand steps a day burns five hundred calories.)

Many people spend the day at work sitting at their desks. There are only so many trips you can take to the mailroom in one day, and so it can be hard to get extra steps into your routine. But there are simple ways to use the environment to reach this goal. When you park in a parking lot, you typically look for the space that is nearest to the entrance of the store where you plan to shop. Instead, park near the entrance to the lot, far away from the stores. That way, you have to walk across the parking lot, which could easily add another five hundred steps to your day.

It is hard to plan in advance for new situations. Instead, when you're working to change your behavior and find yourself in a new place, look at the situation. Figure out where the obstacles are that keep you from achieving your goals. Where are the temptations in the new situation? What can you do to make those temptations difficult to fulfill? Can you remove the temptations from your immediate environment? Can you create extra steps you will have to take before you can give into a temptation?

Are there obstacles that make the actions you want to take hard to perform? Are there things you can do to make those easier to do? A few years ago, I stayed at a Westin Hotel, and they gave me the option of a hotel room with a treadmill and a set of dumbbells in the room. Although the room was a little more expensive, it made it much easier to work out while traveling. Normally, at the end of a long day of work while on the road, it is hard to get motivated to put on workout clothes and to traipse through a hotel in shorts and a T-shirt in search of the four treadmills and a TV set that passes for a fitness center. It was much easier to take advantage of exercise equipment that was already in the room.

You want to create the habit of doing a version of Smart Change parkour. Just as the new devotees to parkour began to see the world differently based on their newly developed ability to move, you want to see the world differently based on your new habits. Learn to identify the obstacles and temptations in your environment and to start to plan for them. Work with others to manipulate the world so that the desirable behaviors are physically easy to perform and the undesirable behaviors are difficult.

The Takeaways

The world governs your behavior more strongly than you might realize. Not only does the mind want to spend as little time thinking about what to do, the body wants to spend as little effort as possible doing it. Without thinking, we often choose the course of action that is easiest in a particular environment.

Often, the easiest action to take is the one that you took last time you were in the same situation. That is the basis of habits. When the environment continues to support your habits, then you keep performing the actions you have taken before. If you want to change your behavior, you need to disrupt the consistent mappings between the world and the undesired behaviors. Then you need to create new consistent mappings that will help you create habits for the new behaviors you want to take.

You also need to use the world to thwart temptations. If you can remove temptations from your environment completely, they'll become less attractive. When you can't remove the temptation, you may still be able to make the temptation hard to do. If you are going to change your environment, though, make sure that you work closely with the people around you so they do not have to engage in the same behavior change as you.

Because you want to take the path of least resistance, you have to become adept at analyzing new environments for potential temptations and other obstacles to achieving your goals. There are often simple ways to alter new environments to make the actions you find desirable easy to do, and the actions that you want to avoid difficult.

SEVEN

engage with others

The Types of Relationships

Relationships and Change

Communication and Neighborhoods

HAVE YOU EVER LIVED IN A REALLY GREAT NEIGH-borhood? The day you moved in, your neighbors started to show up. They each brought a small welcome gift, like a plant or a loaf of banana bread. They probably had recommendations for great local restaurants and stores. They even suggested people to help fix the small things that inevitably go wrong when you move into a new house.

Over time, you found ways to repay your neighbors' kindness. You hosted parties, watered their plants, baby-sat pets or kids, and carried furniture.

Once you get integrated into a good neighborhood, you realize that everyone is looking out for each other. If someone's dog gets loose, then at least one neighbor helps track it down. When a tree limb falls on top of a fence, several people will gather to help lift it off and cut it into pieces.

Not everything functions like your neighborhood. You can borrow a cup of sugar from your neighbor if you find yourself stuck in the middle of a recipe on a Thursday evening, but you can't borrow a cup of sugar from the local grocery store. The store managers may talk about how they want to be a good neighbor, but when it comes right down to it, they just aren't going to lend you anything.

When you're engaged in changing your behavior, you want to be able to rely on the people around you. And that means you are going to have to be a part of a neighborhood.

Types of Relationships

Not long ago, I was talking with a student who had come to my office for a research meeting. When I asked how her day was going, she gave me a quick summary. I was fascinated by how the events of her morning illustrated three of the key types of relationships we engage in.

She had gotten up that morning with the intention of paying her rent. She discovered that her bank balance was dangerously low. In a panic, she called her parents, and her father transferred some money into her account while they were on the phone. Relieved, she wrote a check for her rent and put it in the mail. On her way to school, she popped into a local Starbucks and bought a cup of coffee and a muffin. When she got to the lab, she and another student spent a few hours helping a third student get one of his experiments ready, because a large number of research participants were about to descend on the lab—and graduate students never want to waste an opportunity to collect data.

As simple as this story seems, it illustrates three different kinds of relationships.

The first is the *family*. Family relationships are built on a base of love and obligation. Families spend a lot of time together. Obviously, a nuclear family may live under one roof. But, even the extended family gets together for holidays and to celebrate occasions like birthdays, anniversaries, and other life-cycle events. They swap family stories and reminisce about past events. Families also make frequent phone calls. They send postcards from exotic locations on trips. They reach out to talk just to say hello and to keep up on the latest family news.

One result of all of this interaction is that families also go to great lengths to help each other without really keeping score. When my student's father transferred money into her account, he may have made a comment about being better about saving money, but he ultimately helped out without maintaining an account of how much money he gave. And some day, if he were to get sick or have an injury, I am sure that his daughter will hop on a plane and find ways to get him back on his feet.

Families simply do things for each other. Indeed, most families have a ne'er-do-well relation who is constantly in need of assistance. It seems to make no sense, but the family is constantly bailing this person out, trying to help him get back on his feet. Even though he does not give back to the family as much as he takes, the family continues to provide whatever support they can.

At the other extreme, most people we deal with are *strangers*. You don't know them, and haven't had any significant interactions with them in the past. The basic rules of common courtesy apply in these relationships. You might engage in small talk with the person at the cash register at your local coffee shop, but you are not likely to divulge the deepest details of your life.

Because you have no special relationship with strangers, the transactions you have with them are all based on getting an even trade. You pay for things at store, and you expect to pay an amount of money equal to the value of the goods. Musicians might swap music equipment. When they do, they each expect to get equipment of about the same value in the trade. My student had to get her rent in on time, because her landlord required a particular amount of money to allow her to stay in her apartment. She paid for her coffee at Starbucks, because she and the store treat each other like strangers.

The last kind of relationship is among *neighbors*, who fall somewhere in between family and strangers. Like family, you have a deeper personal relationship with your neighbors than you do with strangers. You have parties with them. You spend time chatting in the yard on weekends. You follow their events of their lives, and they keep track of what is going on with you.

Like families, you do not pay for your services with neighbors. If you were loading a truck at the local hardware store and a stranger came over and spent a lot of time helping you get everything loaded up, you might consider giving them some money to pay for their time. But, if you were unloading a lot of things from your car into a garage and your neighbor came over to help, you would not give him $20 for his work. Instead, you might help him to rake leaves a few weeks later, or offer to babysit his dog. My student helped her fellow students because they are neighbors. At some time in the future, she can expect that other students will assist her when she needs it.

The difference between neighbors and family, though, is that family does not keep score—at least in a healthy family relationship. Parents continue to do things for their children no matter what. Neighbors, on the other hand, may not need to give equal payment in the moment, but they do expect reciprocity. If someone in your neighborhood never chipped in, never hosted any events, never bought Girl Scout cookies, and generally did not return any of the favors given by other neighbors, then that neighbor would no longer be included in activities. The favors would stop. That neighbor would become a stranger.

An easy way to think about these relationships is that there are two correlated trade-offs. The first is the amount of social effort that you invest in a relationship. Strangers require no effort at all, neighbor relationships require much more social effort, and family relationships involve a lot of social investment. The second is the degree of trust in exchanges. Strangers give you very little trust, and so you have to pay for each transaction as it comes along. Neighbors trust that their efforts now will be repaid later, and so they keep track of what they do on a longer and looser schedule. With family, the calculus of exchanges goes out the window altogether.

Relationships and Behavior Change

Your social world has a huge influence on your behavior. When you engage in Smart Change, you need to find ways to use all of the relationships in your world to help you develop new behaviors. By categorizing your world into family, strangers, and neighbors, you can help yourself involve the people around you in behavior change in the most effective way possible.

FAMILY

Family members pose an interesting problem when it comes to behavior change. At times, family can be a huge asset in your attempts to change your behavior. At other times, family is simply a pain (in the asset).

Family relationships are characterized by a willingness to do things for each other without keeping track of what's been done and by a commitment for family members to accept each other as they are.

On the positive side, family members can be counted on to step up to provide lots of assistance and resources to support changes in behavior. In extreme cases in which behavior change is needed (like drug addiction), family members are often the only ones willing to stage an intervention that pushes the addict toward rehab programs and counseling.

Even in less extreme cases, family are often willing to provide help resisting temptations and supporting your achievements as you work toward a contribution. To make that happen, though, you have to be willing to give your family the permission to help you. This can be easier said than done.

In the United States (and other Western countries), there's a strong individualist streak. Western culture prizes independence and self-sufficiency. Our stories of success are almost all focused on great people doing great things in important situations.

Because of the value placed on independence by Western culture, it can be hard to let other people help out with behavior change. It is easy to believe that you are not engaging in true behavior change unless you do it yourself. In the end, your life is not a solo journey. The important thing about Smart Change is whether you end up effectively making your contribution. Getting help along the way does not lessen your accomplishment.

You have to be willing to engage your family and to allow them to help you. If you need time, space, or resources, your family will be most likely to give you what you need without needing much in return.

Family is particularly valuable for providing markers of your progress in behavior change. When you are struggling to make a change, it is hard to recognize the small achievements you have made along the way. In the early stages, when you may be struggling to overcome an old habit or to find ways to engage a new goal, you may not notice what you have done so far. At those times, your family can remind you of your accomplishments to date.

Your family is also in the best position to nag you when you are falling behind. Because small achievements lead to large contributions, it is important to make regular progress. Most people—strangers or neighbors—are not involved deeply enough in your life to see the incremental progress you've made. On those days when you cannot find the energy to engage your Go System on your own, you have to be willing to let your family provide some of that energy for you. You may not enjoy their pushing at the time, but that daily support can make a huge difference as the weeks go by.

Despite the potential positive influence that family may have on behavior change, you may not want to engage all of your family members in your Smart Change efforts.

Family members are often willing to say things to each other that they would never say to friends or strangers. The same repeated criticism from a friend may eventually erode that friendship; you are much less likely to sever your ties with the family member than you are with friends.

But being related does not mean that you necessarily want those people to play a role in your behavior change. If your family members are a source of stress, they may actually get in the way of your ability to resist temptations.

Finally, some of your family members may have a hard time believing you can change. Your family members have spent a lot of time with you. They have a lot of experience with your behaviors, and they may find it difficult to accept that you're committed to making a real change in your life. Just as you will have a hard time changing your behavior if you think it's a fixed trait more than a malleable skill, your family will have difficulty believing you can change if they think the behavior is a core part of who you are. If you are going to bring family into your process of change, you need to be sure that they believe you can change.

Family: How Can (Should) Family Members Be Involved in My Behavior Change?

In your Smart Change Journal, create a page for family. List the family members you see or talk with regularly. For each family member, think honestly about whether you can engage them in your attempts to change your behavior. If you have family members who are not going to be supportive of the work you are doing, then put a line through their name. Even though family can potentially be a great resource, not every member of the family is going to help.

For those family members with whom you do want to engage, it's important to bring them into your plans explicitly. At the beginning of any process of change, it can be hard to be honest with others about your plans. Some changes in behavior reflect aspects of your life that you're self-conscious about. To get the support you need from your family, though, it is important to have a specific conversation with them. Let them know your plans. Let them know how they can be helpful and the things they might want to do that you believe will be *un*helpful.

Most family members want to support you, but they may not know what they can do. An open conversation helps create a positive atmosphere for behavior change.

STRANGERS

The biggest influence of strangers on your behavior comes from goal contagion, or the concept that you tend to adopt the goals of the people around you. When you see other people doing something, it naturally activates that same goal for you.

When you are trying to change your behavior, you need to find people who are pursuing the activities you want to engage in and to avoid the people who are engaged in the behaviors you no longer want to perform.

GOOD STRANGERS	BAD STRANGERS

In your Smart Change Journal, add a page titled "Strangers." Make two columns. At the top of one add the heading "Good Strangers" and at the top of the other add the heading "Bad Strangers."

Under the Good Strangers column, make a list of the kinds of people who engage in the behaviors you want to perform. Where do you find these people? Add places where these people hang out to your list. When you are near the people who engage in the behaviors you want, you will activate that goal.

In Austin, Texas, for example, there is a beautiful trail that skirts the edge of Lady Bird Lake, which cuts through the center of town. From dawn to dusk, there is a steady stream of runners, walkers, and bikers who crowd the trail. There is a public boat dock, and the lake is filled with people on kayaks, sculls, and paddle boards. It is nearly impossible to be near this area without wanting to get moving. Anyone in the Austin area who wants to start an exercise program should hang out by the lake, particularly on days when they feel like they are too tired to do any exercise. A few minutes of watching the activity is enough to engage the goal to join in.

Under the Bad Strangers column, you should make a list of the kinds of people whose goals get in the way of the contribution you want to make. Who are these people, and where do they spend their time? Wherever they are, stay far away. Their motivational energy is poisonous for you, because they will engage your Go System toward the goals you are trying to avoid.

Now, it may seem obvious that you should strive to be near the people who do the things you want to do and to avoid the people who engage in the activities you want to stop. But it still requires effort. Your habits until now have taken you into environments with people who share your interests.

NEIGHBORS

When I talk about neighbors, I refer to the network of people in your life with whom you interact with regularly. Some of these people may be friends. Others can be coworkers. Still others may be people you encounter only in the context of the new behaviors you are developing. And of course, your actual neighbors may be part of this list.

There is evidence that the people closest to you exhibit the same behaviors you do. A fascinating analysis by Nicholas Christakis and James Fowler looked at obesity. They found that people who had lots of social connections who were obese also tended to be obese. Most of your social connections are your neighbors, so you tend to act in similar ways to them.

There are many reasons why your social network can influence behaviors like weight gain and weight loss. Goal contagion is clearly one factor. As previously discussed, it is not unlikely that you will adopt the goals of the people around you just by watching their behavior.

Because your neighbors often determine your physical environment, this environment creates an opportunity for shared behaviors. Your neighbors can just as easily create situations that provide temptations as they can support your desired change.

Your neighbors provide a lot of social support for your actions because the communities to which you belong establish norms for behavior. These norms tell you what you can do within a group. The rules are enforced by the general approval or disapproval of the neighborhood.

When I was a kid in the early 1970s, my father and I went to Indian Guides, a program sponsored by the YMCA. Every month or so, we got together with a group of other kids and their dads to do crafts and activities that were vaguely inspired by Native American traditions. At these meetings, the kids would sit around a table working on some kind of necklace, plaque, or sign that had plastic bear claws, teeth, and feathers on it. The fathers stood around talking, and about half the dads at any given moment were smoking cigarettes.

Thirty years later, I was a father of a son who was in Cub Scouts. The meetings had a similar structure to those Indian Guides meetings of my youth. The boys would be sitting around a table working on some kind of activity. There were no bear claws, teeth, or feathers, but other than that, it was the same. There were still plenty of parents around, and they still stood around chatting while the kids worked.

The big difference is that there was no smoking among the parents. By the year 2000, the social norm had changed completely. It was not appropriate for the parents to smoke around the kids. If someone did want to smoke, he would excuse himself from the group and walk to a place where he could not be seen. This kind of neighborhood norm clearly influences the behavior of the people in the community.

The watchful eyes of neighbors can have a significant influence on behavior simply, in part, because people are much less likely to engage in morally wrong behaviors when there is a chance they will be caught. Religious people have the same experience when thinking about God. Some evidence suggests that many religious people (particularly those who believe that God punishes sins) treat others more fairly and are more likely to engage in moral behavior when they are thinking about God than when they are not.

Who Are the Members of Your Community with Whom You Engage? How Can Your Neighbors Help You Avoid Temptation?

In your Smart Change Journal, list some of your temptations. Think about ways to use your neighbors to help you overcome those temptations. Let your neighbors know what your temptations are.

Then think about ways that you can spend as much time in the company of your neighbors as you can, so your behavior can be observed by others. The more you leave yourself open to the scrutiny of others, the easier it is to help your Stop System prevent you from giving in to the temptation.

As an example, think about the workday. As much as you might want to perform at a peak level for the eight (or so) hours that you are at work, the day itself ebbs and flows. You might be at your best first thing in the morning, then experience a lull around lunchtime, work well for a few hours in the midafternoon, and then wind down as the day ends. In those downtimes, there is a temptation to engage in "fake work," where you sit at your desk looking engaged, but you are really searching random topics on the web or browsing shopping websites. If you feel you are having periods of the day that are not productive, then during those hours make sure that your workspace is visible to your colleagues. You probably do not want your fake work behavior to be visible to others, and so sitting in a public place helps keep you focused on what you need to accomplish.

You do not want the disapproval of your neighbors (or God, for that matter) to be the only reason you are able to avoid a behavior. As I discussed in Chapter 5, the Stop System needs to be used sparingly. The ideal state for behavior change is one in which the Go System focuses you on the desired behaviors. So you want to use your neighbors primarily to reinforce new behavior.

PARTNERS AND MENTORS

It is also valuable to engage with individuals within your community because many goals are much easier to accomplish with a partner or a small group. The Cleveland Clinic offers free yoga classes whenever there are at least four employees at a location who want the class. Yoga provides some physical exercise and stress relief to employees. It is harder for a member of that small community to skip out on a yoga class than it would be if the class were run at a convenient location but attended mostly by strangers.

Alcoholics Anonymous (AA) also understands the value of individual neighbors in behavior change. AA is designed to help alcoholics quit drinking and stay sober by creating a community of people with the common goal of sobriety. The group engages in many community-building practices like holding regular meetings, having social events, and encouraging members to take on a sponsor.

The sponsor is useful in particular because it's easier for the alcoholic to turn away from a community than to turn away from a specific individual. This one-on-one social relationship has a lot of motivational force. It also creates a specific action for the alcoholic to take when suffering from an extreme temptation: Don't think, just call the sponsor and talk.

In Alcoholics Anonymous, the sponsor is a neighbor who has more experience with sobriety and who can serve as a *mentor.*

Who Are Your Potential Guides and Counselors? How Can You Engage Them?

Austin, Texas, has a huge bicycling community. Though the streets are only moderately bike friendly, the weather makes it possible to bike all year long (although in the summer it can be hard to hop on a bike in the middle of the day).

On weekend mornings, there are lots of pairs and small groups of people who are out biking. These exercise partners are usually fairly well matched for ability. After all, it would be frustrating for an expert cyclist with lots of experience and endurance to have to go for a short slow ride with someone just taking up the sport.

Obviously, one reason for cycling with a group is that the bikers can enjoy each other's company. Equally important, though, is that it keeps people engaged with the exercise. On days when a cyclist is busy or tired or too sore to feel like going out, the added force of going with a group can keep the biker on a regular exercise program.

Who Are Your Potential Partners? How Can You Engage Them?

When you start the process of Smart Change, you may not know which people in your neighborhood are the right ones to help you. Don't worry if you draw a blank. Part of the difficulty of behavior change is that there are a lot of unknowns ahead of you. At the start of the process, it is hard to predict what will happen next.

To really figure out with whom you should engage from your neighborhood, you need to become better integrated into the community. And that requires effective communication.

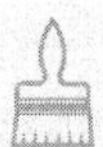

Communication and Community

Our ability to break the world into categories has been a real puzzle for scientists. After all, you see the world quite similarly to the people around you. The animal that lives next door barks at squirrels and wags his tail when he sees you; he's a dog. You know he is a dog, your neighbors know he's a dog, and even the four-year-old across the street knows he is a dog.

This is true for more than just dogs. Pick any random object in your house. Chances are you categorize that object in about the same way as everybody else you know. That is pretty amazing.

How does this happen? Why do you tend to see the world so similarly to the people around you?

For a long time, scientists considered two possible answers to this question. One is that the world really is broken down into categories. That is, there may really be different groups of things in the world and our brains just give us information about what is truly out there.

It certainly feels like that were the case. It just seems obvious that the animal next door is a dog and that nobody could possibly think it was anything other than a dog. We must be responding to what is objectively there in the world.

Yet scientists have also considered a second possibility. Perhaps the world feels as if it were broken up into categories because there is something special about the brain that creates a feeling of order from the chaos around us. Maybe we see the world as having categories, because that's how the human brain operates. You might share your categories with the people around you, because everyone's brains are set up in a similar way.

That possibility seems more plausible when you start looking at other objects in your world. Sit in your kitchen for a while. Look in the cabinet where you keep your plates. Inside that cabinet, there are big plates and small ones. Some of them are saucers that you use for coffee and teacups. You also have some bowls. You might even have a serving tray or two. If you were to put all of these items out on the countertop, they would look rather similar. It can be difficult to figure out exactly where the dividing line is between a large plate and a small serving tray, or between a plate with high sides and a shallow bowl. The world itself has a lot of similarity in it. Yet, you probably agree quite a bit with the people around you about which items are plates and which ones are trays or bowls.

It turns out that both ways of thinking about categories are right to some degree, but both are missing an important piece of the puzzle. There are some categories that do exist in the world. Lions are quite different from other animals, and it is hard to confuse them with anything else. And our brains do seem to have some mechanisms that influence the categories we form. The shape of objects is often more important to the way we form categories than the color or size of the objects, and that probably reflects biases that come from the way our brains are structured.

What is missing, though, is the influence of your community.

From the time you are a young child, the categories you recognize are being shaped by the people around you. Watch a toddler interacting with a parent or preschool teacher. The adult points to an object and gives it a name. "Look at that; it is a ball." The child quickly learns the name for the object and uses that same name for other objects as well. Sometimes, the child uses the wrong name for an object and is corrected. The child might pick up a round beanbag and call it a ball. The adult says, "No, that is a beanbag. Hand it to me."

Over the years, the child gradually learns the way the adults break the world into categories. The process continues throughout life as the child learns new and more specific categories to help him solve more sophisticated problems.

We also learn a lot of new categories when we spend time in new environments. Over the years, I have had the chance to work with many companies, helping them integrate insights from cognitive science into their businesses. Each one has lots of vocabulary to describe concepts that are important to its corporate culture. Nobody teaches us those words explicitly. We never get a cheat sheet for corporate vocabulary. Instead, we pick up on the concepts through conversations with other people.

Without realizing it, we gradually come to think in a similar manner as our neighbors.

Suppose that you and a neighbor get into a discussion about politics. You support different political parties and have very different views about the role of government in the lives of individuals. If you have an extended discussion about a set of issues, you might leave that conversation feeling like you disagree even more than you did when you started. Your views on these contentious political issues may feel even more entrenched.

But the process of having a conversation with this person has changed you both. To understand what your neighbor was saying and to respond to that neighbor's arguments, you had to agree on the words you used to talk. The conversation itself helped refine your categories. You and your neighbor leave that conversation thinking more similarly than you did at the start of the conversation, even if you continue to disagree.

The mutual influence that neighbors can have on the way they think happens only when they communicate with each other. It is the act of having a conversation that creates the change. To discuss something with another person, you have to agree on the meanings of words you use. You resolve your disagreements during the conversation.

That's why it can be so dangerous to disengage from your community. You can end up viewing the world in a radically different way from the people around you. In February 2010, a man named Joseph Stack climbed into his small plane at the regional airport in Georgetown, Texas. He flew several miles south to Austin and crashed his plane into a glass-and-steel office building near a shopping center, killing himself and one man who was in the building, while injuring several others.

It turns out that this crash was no accident. The office building housed a local Internal Revenue Service (IRS) office. Stack had become increasingly angry with the federal government, and on the morning of the crash, he posted a long rambling note about the things that made him angry with the government and the need to pay taxes.

In addition to his career, Stack was also a musician. He played in local bands and at jam sessions where musicians would gather to play together. The musicians he played with were stunned when they heard the news that he had crashed his plane into an IRS building. In all the years that he played with them, he never mentioned his views on the government or the IRS or the anger and distress he was feeling.

Instead, Stack spent his time on websites talking with other people who shared his views. The irony of the Internet is that it allows people who hold extreme opinions to find other like-minded individuals who may live in different places. As a result, communities can develop in the Internet among people who are geographically separated but who hold common views. People need never have a conversation with someone who will disagree with them, and so they will never be forced into a position where they might have to modify their concepts based on the moderating influence of people with less extreme views. Unfortunately, Stack reached the point at which he took violent action on the basis of his increasingly radical views.

This exploration of the way that communicating with others

affects the way you think suggests that communication can play a powerful role in Smart Change. Eventually, you want to reach the stage at which the desired behavior is the one that you carry out by habit. Communication can help with that in a several ways.

Communicating and Optimizing Your Goals

As you engage with new neighbors as a part of behavior change, you'll be introduced to new ways to talk about your goals. These discussions will help you synchronize your goals with other people who have succeeded at changing their behavior. In this way, you learn more effective process goals that allow you to live your life in ways that promote the behaviors you want to continue.

People who attend meetings of Alcoholics Anonymous can feel as though they have stepped into a parallel world. The language used by members of this support program is filled with new words and phrases. Many of the words embody goals that the program uses to help people get beyond their addiction to alcohol. Members frequently talk about *sharing,* which encourages new members to open up about their experiences and to be explicit about their drinking behaviors. They are also asked to create an *inventory* of their behaviors. This language helps new members become mindful of the many habits that have led them to get drunk in the past.

Even the bumper-sticker phrase "One day at a time" has a goal buried in it. In Chapter 5, I talked about the role of psychological distance. When you are distant from a situation, you treat it abstractly, but when you are close up, you think about it specifically.

The phrase *one day at a time* reinforces the idea that a contribution is the combination of lots of achievements. Each day that an alcoholic does not drink is another day on the road toward long-term sobriety. Also, by focusing on the present day, this phrase makes people think specifically about what they are going to do that day to be sober. Additionally, this phrase can help people maximize the effectiveness of their Stop System. When the goal to drink becomes active, this phrase suggests that the alcoholic need only get through the day. They do not need to worry about what will happen tomorrow. By focusing on a narrow time frame, the task of avoiding alcohol may not seem so overwhelming.

Communication also helps you learn more about how to engage with your neighborhood. When you first start on the path to behavior change, you may not really know all of the factors that will make the difference between success and failure. If you look over your Smart Change Journal, you may not be satisfied with your ability to create implementation intentions—those specific plans that determine what you are going to do and when you are going to do it.

As you engage with your neighborhood, you clearly learn more about the goals you need to satisfy to begin to make your contribution. At the same time, you also learn more about your neighbors. You learn about which people are achieving success and which people are struggling with the same aspects of behavior change that you are.

As you get to know your neighbors by talking with them, you can do a better job of selecting potential mentors and potential partners. As I mentioned before, Alcoholics Anonymous encourages members to find sponsors—neighbors with more experience in sobriety who can provide help and support. No matter what contribution you are trying to make, it takes some time to recognize those people in your neighborhood who have the skill to be the kind of mentor you want. Choose your mentors carefully, because they'll have an important influence on your ability to change behavior.

You can also find partners at a similar stage to help you achieve your goals. As you consider potential partners, try to find someone who is as motivated as you are to achieve your goal. You'll draw a lot of energy from your partner. If your partner does not have much energy for change, that will end up hurting your own efforts as well. So although it's valuable to find a partner to help you, it's most important to find a good partner.

It is crucial to really communicate with your neighborhood. A neighborhood is established and maintained by real two-way interactions between people. Communication makes you think similarly to the other people in your neighborhood, and it reinforces the social bond between you. That connection provides you with other people who can help you make your contribution.

And as you journey through behavior change, you should also choose to remain a part of the community. Eventually, you may find you are being called on to mentor others. In Alcoholics Anonymous, everyone who has maintained their sobriety for a period of time is expected to help others, and many of them eventually serve as sponsors themselves.

The value of continuing with the neighborhood is that it helps you think of your behavior change as a process rather than an outcome. As I mentioned in Chapter 3, almost every difficult instance of behavior change is one that requires an ongoing set of achievements to make your contribution. By staying engaged with your neighborhood, you put yourself in a position to treat the change as a process. If you disengage with your neighborhood after you begin to change behavior, you're more likely to think of your goal as an outcome to be achieved. Then when you feel like you have reached that outcome, you might find it difficult to engage in behaviors to maintain it over time.

Relationships and Habits

Changing the people you come into contact with requires changing a set of habits about where you spend your time.

To help you change these habits, you need to track your time a bit more carefully. In your Smart Change Journal, set up a Habit Diary and keep it for at least fourteen days. Questions that you can use to complete your diary include Where did you go today? Were there aspects of the places you went that you think will make it hard—or easy—for you to change? What kinds of people were you engaged with: family, strangers, or neighbors? Did these people make change easier or more difficult? What other issues or factors affected how you were able to work toward your goals? Were you physically or emotionally fit? Was today a holiday or weekend with less stress?

The more you understand about the pattern of your activities, the better able you will be to make changes that bring you into contact with people who will support the goals you want to achieve.

habit diary

Where did I go?
Did the places I go make it harder for me keep on track with my goals?
Who helped make change easier or harder?
What other factors made change easy or difficult?

The Takeaways

There are three broad kinds of relationships you have with people in the world around you: family, strangers, and neighbors.

Family is the set of tight-knit relationships you have—typically with people who are truly a part of your extended family. Family members maintain their relationships by doing lots of activities together. The benefit of family is that they are often willing to do anything for you. However, family can have two downsides. First, they may not push you to change your behavior as strenuously as you need to be pushed. Second, just as family accepts you for who you are, you accept family for who they are. Family members often say things to each other that they would never be able to say to strangers or neighbors (who would no longer want to have anything to do with them). Consequently, family members may not be the best supporters of your attempts to change behavior.

Most people are strangers. You do not know them, and they do not know you. As a result, when you interact with them, you engage in transactions in which each person needs to get something of value in the moment. The main value of strangers in Smart Change is to provide a background of people who may be engaged in similar goals, which can help keep those same goals active for you.

Neighbors are the community of people around you with whom you engage on a regular basis. Neighbors do things for each other without keeping strict score of the value they provide to each other. However, neighbors do feel that the give and take needs to balance out over time.

Neighbors can serve as mentors and partners to provide you with the wisdom of their past experience or are those who share your experience because they are at a similar stage of behavior change.

Communicating with your neighbors creates similarity in the way you all think. This similarity synchronizes the goals of people in a neighborhood. When you immerse yourself in a community of people who have the same goals you wish to achieve, you learn about effective ways to act just by talking with others. You also get a better feel for which members of your neighborhood will serve as good partners and mentors.

Finally, keeping track of your daily environments and interactions with the help of a Habit Diary will help give you a sense of where, who, and what supports your goals for change.

EIGHT

making change

Initiating Your Plan

The Stages of Behavior Change

Self-Compassion

SO FAR, I HAVE ASKED YOU TO DO A LOT OF THINK-ing and precious little doing as you engage in Smart Change. In previous chapters I introduced five sets of tools you can use to attack the most stubborn kinds of behavior change. As I presented these tools, I also encouraged you to keep a Smart Change Journal that lays out your plan.

Optimize Your Goals

You started your journey by learning to optimize your goals. The aim was to figure out both the overall contribution you want to make with your behavior change and the specific achievements you need to engage to turn that contribution into a reality. Chapter 3 encouraged you to develop a set of goals and focus on processes that allow you to change your behavior for the long term.

Tame the Go System

Then you developed a more specific implementation intention in the process of figuring out how to tame the Go System. It's important to recognize that you pursue only the goals that get some degree of arousal, and you can accomplish only specific tasks. By the end of Chapter 4, you had added details to your plan. You figured out the situations in which you were likely to act. You laid out specific actions to take in those situations. You identified the most likely obstacles that would get in the way of making your contribution and planned for what to do when facing them.

Harness the Stop System

Next you focused on ways to control the Stop System, the mechanism that works to keep you from engaging in behaviors that are promoted by the Go System. Although the Stop System is a bad long-term solution to behavior change, it's a crucial part of the overall process. Chapter 5 explored strategies for helping the Stop System work most effectively. In your Smart Change Journal, you added methods for dealing with temptations. Having a specific plan for handling situations that will get in the way of your contribution maximizes your chances of success. I also presented ways of framing your changes to avoid the what-the-hell effect.

Manage Your Environment

Chapter 6 added tools for managing your environment. Your behavior is heavily influenced by the setup of the world around you. In general, you want to minimize the effort you have to put out to act. So the actions that are easiest to take are the ones that are most likely to be taken. In your Smart Change Journal, you examined ways to change the environment around you to make the things you want to do easy and the things you want to avoid difficult.

Engage with People

Finally, Chapter 7 introduced you to ways to use your social environment to change your behavior. You can divide the people in the world up into family, strangers, and neighbors. Each has a role to play in behavior change. Family can provide high levels of support when you need it most. Strangers can influence the goals you adopt through their actions. Most important, by communicating with your neighbors, you can learn new goals and find mentors and partners to assist you. Then you used your Smart Change Journal to begin to identify people who can help you with the changes you want to make in your behavior, recognizing that finding good mentors and partners takes time.

NOW, THE TIME HAS COME TO STOP CONTEMPLATING CHANGE and to start acting on it. That requires a shift from a *thinking* mind-set to a *doing* mind-set. Studies suggest that these two motivational modes are quite different. When you're in the thinking mind-set, you disengage from the world and focus on contemplating what to do next. When you're in the doing mind-set, you are primed for action. In the doing mode, you prepare to engage the world and get things accomplished.

Shifting from the thinking mind-set to the doing mind-set requires immersing yourself into your world. In Chapter 5, I explored the influence of distance on behavior. When you see yourself as mentally far away from a situation, then you adopt a thinking mind-set. When you view yourself as mentally close that situation, you shift to a doing mind-set. To get yourself primed for action, you need to bring yourself closer to your environment.

Implement Your Plan

It is time to start engaging your plan. For many, starting to execute a plan won't be that hard. After all this thinking and planning, it will be a relief to actually start doing something. In fact, it may have been hard for you to keep yourself from getting started because you enjoy doing things more than planning them. But your ultimate success depends on your plan, and so it was good to spend the time working out your path toward the contribution you want to make.

Dealing with Procrastination

Some of you are prone to procrastinate. For you, there is nothing that you can do today that isn't better done tomorrow. If you are a procrastinator, then you need a little extra help pushing yourself into the doing mind-set.

Many procrastinators are low-arousal people, like the ones I mentioned in Chapter 4. Low-arousal people have little motivational energy in their systems on a regular basis. They really need to amp up the importance of a goal to get to work.

If you procrastinate, then you're a person who will benefit from using your environment and your social network to motivate you to begin your journey of Smart Change. For you, the drive to get started is not going to come from inside. You will need to let your environment guide more of your behavior.

Change Your Environment

Start by making changes to your environment. It can be easier to change the environment than to do something that's an overt change to your behavior, because changing your behavior is a direct action, whereas changing your environment feels indirect. However, the changes you make to your environment will start to affect your daily behavior, so those changes are an easy way to get started. If you're thinking of losing weight, for example, start off by removing tempting foods from the kitchen and perhaps by rearranging the plates and silverware to start disrupting your old eating habits.

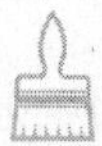

Enlist Your Community

Look at your Smart Change Journal and identify the people you listed as potential mentors and sponsors. Reach out to one of them and get together to talk or spend some time on the phone. Give these people permission to prod you to get started. As with the changes to your environment, talking to someone else can feel less daunting than acting directly. Because your neighbors are an important motivating influence on your behavior, these conversations will quickly lead you to take steps from your implementation intention.

While teaching a class for a company, I met a woman who was about to step into a leadership role for an important project for the first time in her career. She was nervous about this new role as leader, so I encouraged her to find a mentor within the company with whom she could meet regularly to help her get started. Her natural inclination was to take on her responsibilities alone, but having a mentor provided her with an experienced role model who could help her develop a successful leadership plan.

The Commitment Contract

You can get over your procrastination by making the costs of inaction higher than the costs of action. This principle drives the idea of the *commitment contract,* in which you find some neutral person and agree to pay that person money (preferably an amount that would be painful to lose) if you fail to achieve your goal. You might even want to instruct this other person to give the money to a charity that you would not otherwise support if you fail. The goal itself needs to be something that is measurable so that all parties to the contract can agree about whether you succeeded.

This idea is incorporated into the website stickK (stickk.com), which was created by Yale University economists and law professors. The website provides the format for the contract and helps you to get it set up.

As an example, suppose you have decided you want to start exercising regularly. You want to commit to doing three hours of exercise three times a week. With a commitment contract, you would find a neighbor willing to keep track of your progress. You set up a contract saying that you'll pay that neighbor some money. Perhaps you have saved up $750 to buy some new stereo speakers. So you put up that money in the contract. Each week, if you have gone to the gym or walked for at least an hour on three different days, you get to keep your money, otherwise you pay it to your neighbor. You might agree to a few exceptions like being allowed to take days off from exercise if you are sick, but basically you want to make the contract as rigid as possible so you can't keep negotiating new exceptions later.

However, I don't think that commitment contracts alone are a good strategy for changing your behavior. Without a real plan for Smart Change, you're not likely to succeed. The contract will create arousal, but without specific goals that will channel that arousal into action, it will only create stress without moving you toward your goal. You will fail and you will feel bad about that failure in ways that might prevent you from engaging in a contract again in the future.

Once you have a plan, like the one that you laid out in your Smart Change Journal, though, a commitment contract can be helpful. If you have trouble just getting started, then the commitment contract makes the price of doing nothing much higher than the cost of getting started. It can be a powerful tool to help you overcome your general tendency to put things off until later.

Track Your Progress

From the beginning of this journey, you need to start tracking your progress. Your Smart Change Journal can help. From this point forward, your Smart Change Journal becomes a record of your successes and failures. Turn to a new page in your journal and put today's date at the top. At the end of each day, take a couple of minutes to write down how your efforts at change are going. What were your big successes? Did you face any temptations? How did that go? What do you think you could have done better?

tracking your progress

Date:
What were my big successes?
What were the big temptations?
What could I have done better?

Keeping this journal will help you to see which elements of your implementation intention are going well and which ones are not working out as expected. The journal continues the process of helping you become mindful of your behavior as you start to change it.

As you really start to engage in behavior change, though, you also need to think about the way that your pursuit of a contribution will change over time. There is an early, middle, and late stage of Smart Change.

Stages of Change

I liked my college years so much, I never left. For more than twenty-five years, I have lived my life according to the ebb and flow of the semesters. When the students return to school in August, faculty and staff vaguely resent their presence because the entire campus is suddenly swarming with people after a blissful summer with no crowds. After we all adjust to the increase in traffic, though, there is a lot of excitement around. The students are buzzing with energy as they prepare for a new semester. They're all looking forward to what they'll be able to accomplish. They dive into their first few assignments in each class.

By the end of September, that energy is starting to turn into stress. Projects are piling up. An early quiz that goes badly can create some frustration with classes. The great hope that characterized the start of the semester is replaced with the drudgery of the reality of college-level work.

Thanksgiving marks a big change. Right after the break in November, there is a strange combination of panic and hope. On the one hand, lots of classes have big projects and comprehensive exams at the end of the semester, so students' schedules are jam-packed with work. On the other hand, the students also know that in a few short weeks the semester will end, and they will have some time to relax and catch up on their sleep before the process starts again for the spring semester.

The Stages of Excitement

This same pattern holds for behavior change as well. The early stage of behavior change involves a lot of energy and excitement. This is the energy that packs the gym the first week of January as people start to engage their resolution to lose weight in the new year. It's the motivation that drives people to engage energetically with a new volunteer opportunity.

People focused on a particular outcome also start to get energized again as they get near the point at which they are going to reach their goal. Groups trying to raise money for a charity have a lot of enthusiasm as they near the amount of money they want to raise. A runner training for her first 10K race will redouble her efforts as she gets close to being able to run for six miles for the first time. This milestone provides added incentive to keep training.

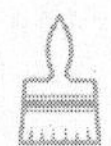

The Stage of Stuck

In between the excitement at the start and the end is the middle, which is probably the most difficult period for behavior change. There are several factors that contribute to making the middle less enjoyable than the start and the end. It is harder to gauge your progress in the middle. Early on, there are lots of achievements. Each visit to the gym allows you to do more than you were able to do before. Every week you play a new musical instrument, there is some obvious new achievement that goes beyond what you could do just a few weeks before.

After a short while, though, it is hard to see the new progress as easily. Psychologists have identified a typical learning curve to describe what happens when people are acquiring a new skill. Think about playing a musical instrument. If you do not know how to play the instrument at all, then the first week that you play you feel a tremendous sense of accomplishment. Suddenly, you go from no ability to play anything to being able to play something—perhaps a simple song. That next week, you learn the same amount that you did the first week, which makes you twice as good about your accomplishments.

Eventually, though, even if you keep learning the same amount each week, it gets harder and harder to notice the weekly improvement. And the reality is probably worse than that. As you get more proficient at a skill, you probably have to work even harder to learn something new, because each additional skill is harder to perfect.

Soon after you start with behavior change, then, your rate of improvement ceases to be rewarding. At that point, many people give up. That's one big reason the gym empties out by the first week of February.

What can you do to avoid being stuck in the middle?

Pay attention to the way you monitor what you are doing. When you start out in pursuit of any goal, you generally pay attention to the progress you have made so far. Because it feels like you accomplish a lot at the start, this perspective can be quite motivating. As soon as your sense of progress begins to slow, though, you have to switch the way you keep track of what you have done.

To get past stuck, change your focus from the starting point to the outcome and start looking at the distance yet to be attained. As you get close to achieving your goal, the focus on what you have yet to accomplish becomes highly motivating. When you first start pursuing a contribution, the distance between where you are and what you hope to achieve may make the journey seem impossible even though you're being motivated by the excitement of the start. As your goal begins to feel attainable, that narrowing of the gap between the present and the future can drive you to work harder. You may be able to overcome the doldrums in the middle by focusing on the distance you still need to cover.

However, at times the distance between where you are and where you want to be is still too large, even though your initial excitement has carried you some long way. In that case, you need to set some intermediate goals to help you continue to progress through the middle. Losing a large amount of weight can be difficult because after the initial joy of modest weight loss, you may have difficulty seeing more progress. At that point, you may want to celebrate each loss of ten pounds on your way to the overall goal. These landmarks provide motivation when the goal itself seems too vast to accomplish.

Many video games are set up like this. When you first play the game, you move up to new levels with new challenges easily. The longer you play, the more you have to accomplish before you move to another level. However, the games always have levels so that there is always some new target to shoot for.

Another potentially good use of the commitment contracts I described earlier is to help you to deal with the middle plateau. If you find your motivation flagging, then a commitment contract can play the role of levels in a video game and provide some landmarks for keeping you going until your broader contribution is in sight.

Engaging Your Community Through the Stages of Change

Your relationship with your neighbors will also change as you move through the stages of Smart Change.

In the early stage, mentors and partners are particularly helpful. They can guide and motivate you as you get started. Mentors help you refine your plan by sharing the wisdom they have gathered from their experiences. They are also able to recognize the signs that you're having trouble overcoming obstacles, so they can provide a lot of guidance early on.

Recently, there has been an increase in the number of people who work as executive coaches. In the modern work environment, many people change jobs several times. It can be daunting to make a career change because it requires a lot of behavior change. Executive coaches provide mentoring by suggesting jobs that fit with a person's skills and developing a résumé that will attract potential employers. You need not seek out executive coaches, though. You can join a professional society in your line of work, through which you will meet many people who have taken the career path you desire. Engage with them and let them help you manage your career.

Partners can help you develop new habits. By working together, both you and your partner will repeat the behaviors often enough to help create a habit. Also, sharing your experience with others who are at the same early stage as you can be valuable in helping you recognize that the difficulties you face are not unique. Working with a partner lets you know you are not alone.

I am a big believer in interacting with colleagues at work to help develop new skills. It's easy to set up a lunch meeting with one or two people who are at a similar career stage to talk about problems, share tips, and get advice. These work partners can support you in a quest to pick up new skills at work. It is also a relief to get to know peers who really understand the problems you are experiencing and who may be able to serve as partners throughout your career.

Indeed, when I go to professional meetings in the cognitive science community, there's a striking phenomenon. You might expect that the groups of people talking with each other at these meetings would be organized around research topics in which people with common interests engage in discussion. Instead, the groups typically consist of people who are about the same age.

Researchers tend to start going to conferences in graduate school, so they meet other grad students and start talking. Soon they find themselves comparing experiences and sharing tips. Over the years, these groups reconvene each year, helping each other through their various career stages.

In the Spirit of Cooperation

One thing that characterizes this early stage is a sense of cooperation. At the start of behavior change, you are prone to reach out to anyone you think might be able to help. You are grateful to hear about other people's experiences and to get suggestions for ways to make behavior change easier. You are also likely to enjoy sharing your successes and giving other people tips to help them change their behavior. It's empowering to feel like you are helping others.

Indeed, one thing to watch out for when you first start changing your behavior is that your conversations with other people may revolve around only the new changes you are making. Because you have to focus so much energy and attention on the new behavior and because you are (justifiably) proud of what you have achieved so far, you can end up dominating conversations with stories about your attempts to change. In that heady period of excitement, try to remember to ask people about how they are doing and what is going on in their world. You don't want to alienate—or bore—those who are most likely to help you.

In the middle stage, your relationship to your neighbors starts to change. As you begin to feel like you're getting the hang of changing your behavior, you may find people at the early stages of change to be annoying. You are no longer that interested in hearing about problems that people who are new to the process are experiencing. You would rather focus on your own progress rather than having to engage as deeply with your neighbors.

In addition, you may shift from feeling cooperative with your partners to feeling competitive. Competition is an interesting reaction.

If you are an actor and are auditioning for the leading role in a play, then you're in direct competition with everyone else auditioning. If someone else gets the part, then you will not. Most aspects of life, though, are not like auditioning for a play. More than one person can succeed. Yet, many people bring this competitive frame to behavior change, particularly in the middle and late stages. When you start to see the people around you as competitors rather than neighbors, you share less information, and you interact with your neighborhood less often. You stop doing the little things that maintain your connection to the people around you.

Behavior Change Is Not a Competition

Resist the urge to treat behavior change like a competition. Even when you are in a competitive market, it is often best to work with your neighbors rather than fight against them. Over the past several years, for example, I have been a blogger for sites like *Psychology Today* and *Huffington Post*. Through this process, I have met a number of other science bloggers and authors. You might expect a group of writers to be competitive. Yet the community of bloggers and authors is actually quite welcoming and mutually supportive. This community has recognized that sharing tips and promoting each other's work increases the success and visibility of the entire group.

This spirit of cooperation is important to maintain, even when you reach the middle and late stages of change. When you find yourself becoming competitive with others, you should work to overcome that urge by mentoring someone else. Sharing information helps reengage you in your own process of behavior change by reminding you how far you have come. It also gives you a chance to be specific about your own behavior. By talking about the way you act now, you may discover things you have begun to do by habit that you would like to refine.

Become a Mentor

Acting as a mentor also keeps you engaged with your neighborhood. The conversations you have with the people you work with help solidify your connections to your neighbors. These connections can be quite important when you reach the doldrums of the middle stage and you need to reenergize your commitment to the goal.

Perhaps the most important lesson to draw from this discussion of stages is that the process of changing your behavior will feel different over time. Strategies you used to motivate yourself early in the process may no longer work as you hit the middle. Neighbors who were incredibly helpful to you when you first started out may feel as though they were getting in your way later on.

A difficult part of dealing with these stages of change is that your progress will not feel continuous. There will be days or weeks when each day is easier than the last. And there will be times when you feel as though you were sliding backward. Some days you may want to give up entirely. And you may even give in to the urge to quit for a while, only to want to try again.

In those moments, you are going to have to learn to give yourself a break.

Self-Compassion

In the 1970s, there were two regions in the United States that stood at the forefront of the technology industry. On the East Coast, in the technology corridor in the Boston area, several large companies like Digital Equipment Corporation (DEC) and Raytheon were at the leading edge of computer technology. These companies made complete computer systems that were used by large companies and universities. At the same time, there was a growing technology community on the West Coast around Stanford University. It was anchored by a few large corporations like Hewlett-Packard, but there were also lots of tiny startup companies.

Over the following twenty years, the fortunes of these two regions diverged. By the end of the 1980s, many of the titans of the computer industry in Boston had shrunk considerably or disappeared altogether. On the West Coast, the technology industry thrived. In the same period that companies like DEC were shrinking, Silicon Valley emerged, churning out companies like Apple and Google, and supporting the growth of the semiconductor industry. Many entrepreneurs found the environment in Silicon Valley to be a hospitable place to start a business. There were lots of talented technologists and experienced business professionals and bankers to support any group with a good idea and the interest in developing it into a company. Stanford University opened its doors to companies, and many students got involved in the business activities in the region.

Economic history is complicated, and there are lots of reasons Silicon Valley emerged as the technology winner in the 1980s and 1990s, while Boston lagged behind. Because of the importance of Boston and Silicon Valley in the high-tech industries in the United States, a number of sociologists and historians have examined these regions. The idea is that if we understand the factors that led to the success in Silicon Valley, then we could duplicate that success in other places around the world that aim to become powerhouses in the technology industry.

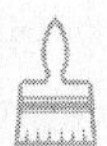

Tolerance for Failure

It turns out that one of the big differences between these regions was psychological—there was a much greater tolerance for failure on the West Coast than on the East Coast. The large East Coast companies were averse to failure. An executive who spearheaded a project that didn't work out well was unlikely to get promoted. As a result, the large companies were often conservative in the projects they took on.

Because of their fear of failure, these large companies were also insular. They did not generally want to share information with other companies. By keeping their new ventures secret, the companies felt they were maximizing their competitive advantage. In addition, if few people outside a company knew the details about a new project and that project never made it to market, the rest of the technology world would be unaware of the failure.

However, by playing their cards close to the vest, the East Coast companies also failed to create a neighborhood. There was little movement of people or ideas from one company to another. This posed a problem, because the technology world was rapidly evolving. In the 1970s, most computers were large. A company or university would purchase a room-size computer system and then have various employees access that system from terminals at their desk. By the 1990s, though, everyone had their own computer at their desk, and networking these smaller machines was a critical part of a company's information technology structure.

Without a neighborhood, the larger companies found it difficult to change their behavior. Long after the business community had started to switch to smaller desktop computers, the big East Coast companies were still focused on providing large-scale computing solutions. They were slow to embrace the changes sweeping through the industry.

A markedly different attitude held in Silicon Valley. Small companies would form around new ideas. Many of these companies would not make entire computer systems, but rather would focus on specific components like hard drives, specialized computer chips, or networking machinery. Some of these ideas turned out to be great successes, whereas others were equally spectacular failures. But, unlike on the East Coast, these failures were not harmful to people's careers. Someone who ran a startup that failed would soon end up at the head of another venture. On the West Coast, people wore their failures as a badge of pride rather than as a black mark.

The willingness to fail was one factor that supported a more open environment. Companies still wanted to succeed, and there was still competition among companies, but it was much more common for companies to share information. Companies sometimes even formed what are called "co-opetition" relationships, in which they worked for a short period of time with a competitor to make sure a project succeeded. This flow of discussion, ideas, and money created a neighborhood that helped different companies see the emerging trends in the field and to capitalize on them.

Because Silicon Valley companies were allowed to fail without ruining the careers of the people involved, entrepreneurs in this region were willing to take risks and try new ideas. They learned the key distinction between failure and negligence. A group that works together at peak capacity and tries to turn an idea into a successful reality may fail, but they did everything they could to make it succeed. In a deep sense, that is not a failure but a learning experience. The only true failure is not to put in a good effort and not to do everything possible to make a project succeed.

Learning How to Fail

One reason the tolerance for failure is so striking is that we do not teach people how to fail in our education system. From the earliest grades, our schools are focused on success. The purpose of exams is to get questions correct. The people who are rewarded in school are the ones who get the best grades, not the ones who take the biggest risks or the ones who learn from their mistakes.

Eventually, though, everyone fails. It is just not possible to do anything new and interesting without failing at least some of the time. It is the ability to learn from a failure and to do better that is the key to success in any venture, including Smart Change. The scientist and poet Piet Hein wrote short verses that he called *grooks*. One of his most famous grooks was called "The Road to Wisdom," and it reads:

Well, it's plain
and simple to express.
Err and err and err again,
but less and less and less.

As this verse suggests, the key to wisdom is not succeeding, but learning from your mistakes in ways that allow the mistakes you make in the future less damaging than the ones you made in the past.

You need to take the proper orientation toward failure. You have to learn the art of self-compassion—treating yourself with warmth and understanding. You can set high expectations, but you should not punish yourself when you fail.

This concept might sound similar to the related idea of self-esteem. Self-esteem refers to the ability to think positively about yourself. It is important for people to think that they are valuable and to believe that they have the capacity to contribute to the world. Self-esteem helps give people the confidence to present themselves in public settings and to make sure that they are treated fairly.

But self-compassion goes beyond self-esteem because it is focused on your response to failure. It relates to your beliefs about what failure tells you about yourself.

Back in Chapter 5, I talked about the work of Carol Dweck and her colleagues whose research explored the difference between two kinds of mind-sets. The entity mind-set assumes that a particular characteristic is an unchangeable part of who you are. The incremental mind-set assumes that some aspect of yourself can be changed with enough effort.

Self-compassion involves taking an incremental mind-set about the source of your failure. If you have a high-degree of self-compassion, you look at your failures and believe they reflect a combination of the actions you took, the situation that occurred, and other factors that may have been out of your control. The key, though, is that you recognize that you can change your own behavior in the future and lower the probability of failing again.

If you don't have self-compassion, then you're taking an entity mind-set about failure. When you fail, you assume that failure is telling you something fundamental about yourself. The failure points out your limits. If you come to believe that failures are things that you cannot overcome, then your response to failure is to give up. Without self-compassion, you start to accept that there are some changes you cannot make.

Self-compassion is the last key piece in our Smart Change puzzle.

Because you're reading this book, there is a good chance you have failed in some of your attempts to change your behavior in the past. You may even believe you are not the sort of person who can change.

The purpose of this book is to help you understand the way your action system works. The reason some behaviors are so hard to change is that your motivational system is exquisitely organized to help you achieve your goals. This system wants you to be able to act without thinking as much as possible. Making radical changes to your behavior is hard precisely because this system is so effective.

But your motivational system learned the behaviors you're trying to change. And so it can also learn the new behaviors you're trying to incorporate into your life. Your motivational system can and will adapt to the new behaviors you want to create, though it will take time.

The five sets of tools described earlier and reiterated at the top of this chapter are aimed at the pressure points in your motivational system. These tools are designed to help you use the structure of your motivational system to your advantage as you change your behavior. The reason I asked you to spend so much time planning your behavior change is because of how difficult it can be to make changes.

As you go through your process of Smart Change, there will be days you will fail. You will eat too much, smoke a cigarette, skip your homework, avoid practicing your instrument, or get angry at a coworker. You have to meet those failures with self-compassion. Failing on a particular day is not a sign that you cannot change. It's a sign that your motivational system is still being efficient at promoting the behaviors you want to change.

But you have all of the tools to make the contribution you ultimately want. A few small failures may indicate that you need some more time to let the process of change reprogram your Go System. In that case, be patient and let your plan unfold.

If you fail systematically, though, then you want to go back through your Smart Change Journal and start to diagnose the problem. What are the situations that are causing you to fail? Where are you when that happens? With whom are you spending time with?

As you begin to understand where these failures occur, think about the tools you can use to help you get beyond those failures. Do you need to reorient your goals to create a different set of achievements? Are there steps missing in your implementation intention that require you to revise your plan? Are there situations in which you are trying to rely too heavily on willpower to get yourself beyond temptations? Are there aspects of your environment that are pushing you toward behaviors you would prefer to avoid? Are there people in your neighborhood whom you ought to engage to help you act differently?

Self-compassion means accepting that failure is a signal you need to do some more work. So trust in the process. In the end, you *can* change your behavior.

Get to work.

The Takeaways

You have to flip yourself from a thinking mind-set to a doing mind-set. To make that happen, start to engage yourself in the world. Bring yourself mentally nearer to the behaviors you want to change. If you are someone who prefers an action mind-set, then you have probably been champing at the bit to get started for a while now. Be cautious, however, and make sure your plan for change is well established. If you are more prone to procrastination, though, then you may need some help getting started. Make changes in your environment and engage your neighbors to allow you to start doing things to initiate changes in your behavior.

Once you get started changing your behavior, you will discover the process of change is dynamic. The strategies that worked in the first week of your new program of Smart Change may not be so effective a month later or even a year later. Behavior change is a process that unfolds over time.

At the early stage of behavior change, it is often motivating enough just to focus on the progress you have made so far. You can also get a lot out of the mentors and partners you find in your neighborhood.

Soon, though, it becomes hard to see your progress. You are probably still moving forward toward your contribution, but it can be difficult to see the changes. In the middle stage, create some landmarks to provide continued motivation. Consider using a commitment contract if you are having trouble sticking to the plan. Engage your neighbors to continue implementing your plan. Start looking toward the outcomes you desire. As those outcomes get closer, they may be more motivating to you than the distance you have traveled since you started changing your behavior.

Resist the temptation to disengage with your neighbors and to treat them as competitors. Instead, consider becoming a mentor to others. Share your experience and continue the conversation with your neighbors. That will help you understand your own process of change.

Finally, treat yourself with compassion. The old adage says that change involves two steps forward and one step back. On the days when you feel as if you have taken a step back, remember these little failures are not telling you that change is impossible. They may be signals that your plan needs to be revised. Keep track of your successes and failures in your Smart Change Journal and use that information to help you think about ways to use the tools for change in new ways to help you overcome the obstacles you face.

NINE

smart change for other people

Changing Other People

Trust and Authenticity

Eight Ways to Influence Others

THE MODERN WORLD IS AWASH IN NEAR-CONSTANT persuasion. You are surrounded by advertisements for products that want to change your goals, aspirations, purchases, and actions. The media are full of stories about new advances and opportunities that could lead you to live your life differently. Your work life is driven by structures that aim to help you maximize the contribution you make to the organization you work for. Your government implements policies that influence your behaviors by affecting what is easy to do and what is difficult.

In short, although your life is your own, there are a lot of people who want to control your behavior.

And most of them are going about it the wrong way.

True change requires bringing a variety of the tools of Smart Change to bear on a person's behavior. But most groups that are trying to influence you tend to focus on just one of those tools. It is rare that anyone develops a concerted campaign to affect your actions.

Most advertising hopes to attract your attention, at least for a short period of time. Often, ads present some kind of information aimed to convince you that a specific product is superior. An ad might tell you about the cleaning power of a new detergent or a special material in a diaper that allows it to be more absorbent. It might show people using the product to give you an idea about how the product might fit into your life. The ad might feature a celebrity spokesperson to add a little star power to enhance the message.

Almost all of these ads are focused primarily on your goals. They want to help you create new goals and give arousal to goals you already have. The assumption behind this kind of advertising is that if you can create a goal (or a need) in a potential customer, then the customer will act to achieve that goal. It should be clear by now, though, that just setting a goal is not really enough to change behavior.

Other attempts at persuasion focus on the environment. Go to a big-box retailer like Walmart or Target. These stores use the layout of the facility to influence what you purchase. The pathways through the store are organized to make you walk past lots of items that you might not ordinarily think about buying. The shelves at the end of each aisle (called the *endcaps*) are stocked with products the store wants you to see. Some stores even sell the most desirable real estate in the store to companies who want their products to get preferential treatment. By putting you in the vicinity of these items, the store hopes you'll reach out and grab a few and put them in your cart.

And you probably will. Because the environment has a big influence on the way you act. But the next time you go to the store, the items on the endcaps and at the front of the store will be different. So the store is not really creating a long-term change in your behavior. Instead, the store (along with the companies that pay for good locations) is assuming that if you buy a product once you can be turned into a loyal customer and you will now go out of your way to look for that product in the future. But if you have to look in a different place for a product every time you go into the store, you cannot develop a habit to buy it.

It is rare to find a group like Dr. Roizen's at the Cleveland Clinic that uses many different tools to affect people's behavior. The Cleveland Clinic wants to change the health-related behaviors of its employees and patients, so it uses many of the tools of Smart Change. I have mentioned several of the approaches used by the Cleveland Clinic throughout the book, but I want to highlight them again here. A serious attempt to help people change their behavior requires creating a lot of structures for people (and also spending some money up front to develop those structures).

To encourage people to take better care of their health, the Cleveland Clinic tries to make the goal of being healthy attractive. One way they do that is to change the costs related to employees' behavior. Employees who engage in healthy behaviors pay less for their health insurance than those who do not take care of their health. This incentive is enough to activate the goal to stay healthy.

The clinic helps people change their habits by easing the process of creating and carrying out plans. They provide yoga classes to groups of four or more employees who want them. In this way, groups can get into a habit of regular movement and stress relief.

The clinic uses the environment to help remove temptations and minimize the demands on the Stop System. The entire Cleveland Clinic is smoke free, so cigarette smokers may need to walk several blocks if they want to smoke. There are no sugary drinks available in vending machines and cafeterias, so such empty calories are not an easily available option. The clinic sponsors farmers' markets on its campuses to make it easier for people to buy fresh produce.

And the Cleveland Clinic has created a neighborhood of wellness by sponsoring employee walking groups and fitness competitions to create partnerships among employees who want to get in better shape. There is also a network of counselors that employees can call or email regularly to get feedback on their eating and exercise habits. So the clinic is providing mentors to help employees through the early stages of behavior change.

The Cleveland Clinic has been successful at reducing its healthcare costs by engaging in a comprehensive program of Smart Change for its staff. Any one of these programs alone probably would not have been successful. By using all of the tools of Smart Change, though, they have been able to change the behavior of a large number of people and have begun to reduce their level of healthcare spending. This program was not free. They invested money at the start with the aim of saving money later.

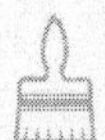

Engaging in People's Lives to Effect Change

Like the Cleveland Clinic, many individuals, organizations, and companies may be trying to change the behavior of the people in their sphere of influence. And to be successful, they need to get people to interact with them on a regular basis.

Volunteer organizations have the same problem. Many nonprofits need a small army of volunteers to help them accomplish their goal. Austin, Texas, has adopted the goal to be a no-kill animal city. In many cities, abandoned pets that are brought to a shelter are immediately evaluated for their potential to be adopted. The young, healthy dogs in the most popular breeds are put up for adoption, but older dogs and dogs from less popular breeds (like pit bulls) are euthanized.

To minimize the number of dogs and cats that are euthanized in Austin, several nonprofits have sprung up to fill the gap. Groups like Austin Pets Alive! (APA) take animals from the city shelters and give them a temporary home until they can be adopted by a family. They encourage volunteers to serve as foster owners to dogs until a family takes them into their home permanently. Of course, there are lots of "foster failures" in which a family that takes a dog into its home temporarily ends up becoming its adoptive parents.

Austin Pets Alive! has some employees, such as office managers, veterinarians, and trainers, but they rely on an army of volunteers. People go to the shelter to walk the dogs, bathe them, and take care of dogs that are scared or aggressive.

Like many organizations, APA has trouble keeping volunteers active in the group. Although large numbers of people attend their orientations, few of them go on to volunteer on a regular basis.

But Austin Pets Alive! is doing an excellent job of creating a neighborhood to help keep people engaged with the group. Using social media like Facebook, APA encourages new volunteers to ask questions and to write about the pets they interact with. More experienced volunteers answer questions for new members. Each day, the office manager posts a list of the pets that were adopted that day to help keep volunteers motivated to work. The organization has even created an explicit mentoring system in which long-time volunteers work with people who attend the orientation to integrate them into the APA community.

Still, they lose a lot of potential volunteers. The problem is that the methods APA uses are haphazard—like those of so many companies and volunteer organizations that want to affect people's behavior. By following the principles of Smart Change, they could help turn more occasional volunteers into committed assistants.

Changing Your Own Behavior to Affect Others

The tools of Smart Change can be combined to have a powerful influence on other people's behavior.

To understand how this process works, it's valuable to think about the ways other people try to influence your behavior. Because so much of your behavior is driven by habits, there are many actions you take on a daily basis that you do not choose to take. To the extent that other people are affecting your environment, your neighborhood, and the development of your habits, you may have ceded control of your behavior to those individuals. Understanding the ways that people can manipulate your motivational system will allow you to recognize when your actions are being affected by others. So even if you have no interest in influencing other people's behavior, the tools explored in this chapter may still teach you about yourself and your environment.

Before drawing some lessons from the five sets of tools I presented, though, it is important to talk about the role of trust when affecting people's behavior.

Trust and Authenticity

When I was in fifth grade, Bubble Yum was *the* gum to chew. It came in little bricks of soft gum with better flavors than Juicy Fruit or Bazooka. If you were going to be one of the cool kids in grade school, you had to chew Bubble Yum.

And then the news hit.

"They" had discovered spider eggs in Bubble Yum. We never knew who "they" were, but at that age, we assumed there was a super-secret government agency that was constantly testing the candy supply for safety.

Suddenly, nobody wanted to be anywhere near Bubble Yum. Bazooka made a comeback among my friends. Several weeks after we all stopped chewing Bubble Yum, there was a story in the newspaper that the entire thing was a hoax. Someone had started spreading rumors about Bubble Yum, but it wasn't true. The company itself took out ads in the newspaper to try to quell the rumors.

The damage was already done, though. None of my friends wanted to go back to Bubble Yum. The gum had been tainted by the rumor, even if it was false. Eventually, Bubble Yum as a brand recovered, but their sales didn't come from anyone I knew. Another unsuspecting batch of fifth graders was probably responsible for the comeback.

This episode says a lot about the way trust and belief work. Because you learn so much about the world from the people around you, you're wired to believe what you hear. It is hard to go through life being skeptical of everything you're told because there is so much to learn out there. And it's impossible to verify everything you hear for yourself. So if you think the source of the information is trustworthy, then you add the new knowledge to your network of beliefs and move on.

Once that knowledge is in your system, it is remarkably difficult to remove. There is no mechanism in the brain that allows you just to delete something that you have discovered is false, at least not in the way you might delete a bad sentence from a document in a word processor. Instead, you have to "mark" that information as being incorrect. But the markers you place in your memory are generally pretty ineffective, and so people often continue to give some credence to rumors long after they have been disproved. It is not uncommon for us to hold on to the axiom, "There must be some truth to it if so many people have heard the rumor."

If you come to believe that a particular individual or a product is untrustworthy, though, that eventually has a significant impact on your behavior. When you trust a product, you are willing to use it without thinking. No fifth grader would think twice about accepting a piece of Bubble Yum before the rumor started. When you have reason to mistrust the product, though, it engages your Stop System. It shifts you from a doing mode to a thinking mode. Once you enter into that thinking mode, you are prone to avoid action.

A lack of trust has the same effect on your behavior as other negative emotions like disgust. If you offer a piece of chocolate to a candy lover, she will gladly take it. But, if you offer that same person a piece of chocolate in the shape of a cockroach, she is much less likely to bite into it. Even if she is certain it is a piece of chocolate, she no longer trusts it, and so she switches from a doing mind-set to a thinking mind-set.

If you want to influence other people's behavior, then you need to develop trust. Without trust, you are putting people in a position in which they are going to enter a thinking mind-set rather than a doing mind-set every time they interact with you. That mind-set is not conducive to getting people to change their behavior.

The core of trust in persuasive interactions is authenticity. Authenticity is the degree to which people think that the public face you have adopted fits who you really are inside. When people feel you are telling them things you truly believe, they are less likely to be skeptical of their interactions with you.

When I was a kid, I used to watch New York Mets games on Channel 9. One of the main sponsors of the games in that era was Schaefer Beer. They had a catchy jingle with the refrain, "Schaefer is the one beer to have when you're having more than one." As a kid, I don't think I took in the full import of that slogan, though later I saw it as refreshing that a beer company would market itself as the beer you should drink if you were planning to get hammered.

Starting in the 1980s, public opinion about drinking—and particularly about drinking and driving—began to shift. Groups like Mothers Against Drunk Driving (MADD) had a profound effect on attitudes toward driving while under the influence. And the beer companies began to respond to this change with notes on their ads inviting their consumers to drink responsibly. This tagline was probably added by breweries as a way to protect themselves from the claim that they were encouraging heavy drinking that could ultimately lead to lawsuits.

That tagline has always rung hollow with me (and many other people as well, judging from the comments on discussion boards about this topic on the Internet). Beer brewers are in the business of selling alcohol, and some of their patrons are going to get drunk. That is probably good for business. For these companies to act as though that were not their intent is not authentic. At a minimum, people will tune out this warning because they are skeptical of it. Worse yet for the advertisers, it may cause some people to ignore the entire message of the ad.

By contrast, the Cleveland Clinic is acting authentically when it tries to help others change their health-related behaviors. They are a large healthcare company, and so when they create wellness programs for their employees, they are working squarely within their mission. Even though reducing healthcare costs clearly helps the company be more profitable, their efforts are not viewed simply as a way of making more money.

Before you hope to establish trust and influence others, you have to see yourself as others see you. What do people perceive your motivations to be? What do people believe is your core area of expertise? You may not have a clear sense of how you are viewed by others, so it can be valuable to gather some information. If you are engaging in this process as part of a company or volunteer organization, then ask questions of your customers to get a handle on how they think about you. If you are acting as an individual, then make sure the things you want other people to change about themselves are things you do effectively yourself. "Do as I say and not as I do," is not a formula for success.

Once you understand your public face, you need to focus your communications around messages that are consistent with the way you are viewed by others. The aim is to avoid having people adopt a skeptical attitude when interacting with you. Behavior change is hard enough to accomplish when people are willing to engage in the process. When they have reason to shy away from this process because they are concerned about your motives, then you have made the process of helping other people change their behavior even more challenging.

Eight Ways to Help Others Change

While any one of the suggestions provided here will work to some degree when tried alone, combining them is more effective than using them in isolation. Therefore, it is important to develop a comprehensive plan to address all of the pressure points in people's motivational systems.

1. Lead by example. Earlier in the book I talked about *goal contagion*. The goals you adopt are influenced by the people around you. When you are trying to change your behavior, it is valuable to be surrounded by other people who are pursuing the same ultimate goal as you are—that is, the contribution you want to make.

When you want other people to change their behavior, it is important to engage visibly in the goals that you want them to adopt. When you lead by example, your actions will serve as a source of goal contagion for other people in your environment. Your actions help people see how a goal can be accomplished successfully. A good implementation intention is a key part of making a contribution.

Many professors I know make a point of working in their offices with their doors open. It would probably be more productive for them to keep their doors closed or perhaps even to work from home. They would get more done if they were interrupted less often. The point of working in a visible way, though, is to give students in the department a sense of what is required for success in academia. To balance teaching classes, doing research, writing papers, reviewing papers for journals, writing grants, and doing administrative service for the university, most faculty I know put in long workdays. Faculty talk to their students about the importance of putting in this time, but it is easier for students to internalize the effort needed to succeed by seeing their faculty mentors at work.

Leading by example is the best way to demonstrate authenticity. You're going to be evaluated by other people based on whether your message is consistent with who you are perceived to be. If you engage in the same behaviors you expect of others, then you're exhibiting the highest form of authenticity. You practice what you preach.

2. Suggest goals. Most businesses would never survive if they sold a product to a person only once. Instead, to stay in business, they need their customers to purchase from them repeatedly over the course of their lives. To make that happen, companies need to make sure their products and services are used regularly, and that means that they need to do an effective job of recommending goals and giving people the motivational energy to satisfy them.

People's actions are driven by specific circumstances. If you show people the conditions in which the product is used, then they will be reminded to use the product in those situations when they encounter them later. In essence, you are helping people develop an implementation intention.

Creating goals in customers is one reason product placements are valuable in TV shows and movies. Companies pay production companies to include their product in scenes. These placements are most noticeable when car companies pay for their cars to be used throughout a film. But it may be even more effective for companies to pay to have characters use their products in scenes of daily life to demonstrate how the product can be integrated into a routine. The key is to find several ways to help potential users of a product or service integrate it into their lives.

Another way to suggest a goal is to get community leaders to engage in a behavior in a visible way. Nonprofits like to publicize when celebrities and civic leaders participate in their events as a way to encourage others to adopt the same goal. In addition, sponsorships can help promote goals. Many musicians that I know get deals with clothing companies to wear their products and to give out cards that provide discounts for fans.

3. Give the right feedback. How do you encourage other people when they are changing their behavior? Chances are you want to be supportive, so you focus on saying positive things to people. And that can be great. But there are several potential problems that lurk when you give feedback to others.

The comments you make to other people affect the way that they characterize their goals. It's common to want to talk to people about the progress they're making toward a goal they're working on. However, there is a danger that if most of your discussion focuses on progress toward reaching some end state, then people may develop outcome goals rather than process goals.

When you see a friend who is on a diet and has been losing a lot of weight, it's tempting to tell her that she look greats and she must feel wonderful. It feels good for someone to hear positive comments, and this feedback will often be encouraging. However, if you end the discussion there, then the only feedback your friend is getting is about her progress toward an outcome. Instead, continue the discussion. Ask about what she is doing that has allowed her to be so successful. What is she eating? Where is she working out? What are the lifestyle changes she has made? When the conversation focuses on the *process of change* rather than the outcome, it reinforces the value of creating a sustainable process whose side effect is the desired long-term contribution.

In addition, feedback can influence the mind-set people adopt about behavior and motivation. Several times in this book, I have talked about the entity and incremental mind-sets. People often give others feedback that inadvertently reinforces an entity mind-set. If you see a friend on a diet at a party eating a small plate of fruit, you might say to him, "Wow, you have remarkable willpower, I couldn't do that." On the surface, this is a compliment. However, underlying this statement is the idea that willpower is an entity that cannot be changed. The dieter might be exhibiting great willpower in that circumstance, but if he gives in to temptation in some other circumstance, does that now mean that he has now reached the limits of his willpower?

It is better to give positive feedback that does not reinforce an entity mind-set. For that same dieter, you say, "I'm impressed that you have managed to avoid all of these tempting desserts. What is your secret?" You are still providing a positive message, but you are not assuming that there is some fixed capacity for willpower. Instead, you're inviting him to tell you about all of the strategies he has put together to support his success at sticking to his diet under difficult circumstances. This kind of feedback promotes an incremental mind-set, which acknowledges that most abilities are skills that can be nurtured.

Finally, the encouragement you give needs to be tailored to a person's stage of change. Research by Ayelet Fishbach and her colleagues at the University of Chicago shows that positive and negative feedback have different influences on people. Positive feedback helps make people more committed to a goal. Negative feedback is particularly good for spurring people to make more progress.

When people are first starting to change their behavior, positive feedback is valuable because it helps people feel a greater sense of commitment toward the goal they want to achieve. These early stages of behavior change can be a fragile time, so it is good to reinforce commitment to change. Over time, however, people shift their own thinking away from their overall commitment to the goal to their sense of progress. At that point, they are motivated by negative feedback, which reminds them of the distance between where they are now and where they would like to be.

Of course, this negative feedback does not make people feel good. Even in the later stages of behavior change, people still enjoy getting positive feedback more than they enjoy getting negative feedback. But at the later stages of change, the positive feedback is not nearly as motivating as the negative feedback.

Although it can be difficult to give people negative feedback, it is important to be willing to make people uncomfortable when working with them to change behavior. If you're helping people to manage their careers, then you can use discomfort to help them get motivated to seek a promotion. Studies suggest that when you focus people on the contribution they have made at work, they are happy with their current job but they do not actively seek a promotion. If you focus people on what still remains to be achieved in their careers, then they feel bad about their current job but are motivated to move upward.

To give people negative feedback, though, you have to be willing to overcome your natural tendency to be *agreeable.* Agreeableness is one of the five basic personality dimensions, and it reflects how much you want other people to like you. All of us are agreeable to some extent. The more you want people to like you, though, the harder you find it to give people negative feedback because in that moment they do not like you so much. Remind yourself that giving negative feedback to people who are already committed to behavior change can spur them to improve and to advance in their careers. So even though it may be difficult to be the bearer of bad news, it is also important.

4. Support habit development. Habits emerge whenever there is a consistent mapping between the world and a behavior *and* repeating that behavior in a specific situation. When you are trying to influence the behavior of other people, you can affect both elements of this formula.

In his book *The Checklist Manifesto,* surgeon Atul Gawande extols the virtues of checklists in a variety of situations in which the same task has to be performed repeatedly. He talks about how one significant source of infections in hospitals comes when a staff member in the intensive care unit (ICU) has to put in a central line, which is a long thin tube that's inserted into a vein in the chest so medicines can be delivered directly into the bloodstream. When these lines get infected, it can put ICU patients (who are already quite sick) in serious danger.

As Gawande points out, if the ICU staff covers the patient with a drape when the line is being inserted and uses chlorhexidine soap, then the incidence of these infections goes down dramatically. Hospitals in Michigan got a medical equipment manufacturer to bundle the drapes and the soap in a single kit and then gave staff in the ICUs a checklist to make sure that they carried out each step in the same order every time it was done. This combination of changes to the environment and routine created a consistent mapping that was repeated often. This process lowered the incidence of central line infections to near zero, which greatly improved patient outcomes.

When you want to change the behavior of the people around you, think about how you can create consistent mappings in the environment. Are there methods of getting people to reorganize their environment in ways that will support the creation of habits? Can you influence people to perform an action often enough that the Go System will acquire a habit?

One place where thinking about habits is particularly useful is in user interface design. People quickly develop habits for the computer programs, websites, and devices they use frequently. These habits allow them to perform actions they engage in a lot without having to think about them.

It is important to keep the basic structure of the user interface for products as consistent as possible over time. Otherwise, they run the risk of disrupting the habits of their customers.

As one example, the social media site Facebook managed to snare almost everyone into their network. In the period from 2009 to 2012, the site was constantly trying to improve its service, and so every once in a while, they would release a new user interface. The hope was to improve the experience of users. Yet, each time that the interface was changed, there was a huge outcry from users because their previous habits no longer worked. The changes in the design were typically inconsistent with what people had been doing before. Although Facebook wanted to improve the user experience, the disruptions they caused to people's habits made people feel bad about the changes.

By contrast, Amazon employs a team to do user interface testing. They know exactly which features of the website people are using to search for products, to look for product-related information, and to make purchases. They are constantly making updates to their user interface, but few customers even notice the changes, because the elements that customers use by habit are kept the same. In this way, Amazon is able to improve the customer experience without disrupting the habits of frequent shoppers.

If you want to influence people's behavior, you need to understand their habits and to work with them. Create consistent mappings within their environment and stick with those mappings. Then give people the opportunity to repeat behaviors in that environment to ensure they turn their actions into habits.

5. Take advantage of laziness. Laziness is a key factor that maintains behavior. People want to minimize both the amount of time that has to be spent thinking about their behavior and the amount of effort required to act. You want to make the behaviors in which you wish people to engage as easy as possible to perform and to make undesirable behaviors hard to perform.

The simplest way to make this happen is to have control over people's environment. The Cleveland Clinic discouraged smoking by making their campuses smoke free. As a result, smokers had to walk a long way just to have a cigarette. So cigarette smoking was hard to do.

Environments can also be manipulated to encourage desired behaviors. The city of Austin has installed a number of dog hygiene stations all over town. These stations consist of a garbage can with a liner and a dispenser with plastic mitts that can be used to pick up dog waste. These stations make it easier for dog owners to clean up after their dogs, which cuts down on the number of people who fail to do so.

Many cities want to encourage more residents to ride bicycles instead of using their cars in the central parts of the city. To promote this behavior, they have set up bicycle-sharing systems. In Tel Aviv and New York City, for example, there are kiosks where members of the bicycle-sharing system can insert a card and take a bicycle. The bikes can be dropped off at any of the other kiosks around the city. The designers of these systems made an effort to locate the kiosks every few blocks to make it convenient for people to use the bicycles frequently. Putting bicycles in the environment makes it more attractive for residents to bike around the city than it is to drive or take a taxi.

You may not always have direct control over people's environments, but you can use the design of products and packaging to encourage people to structure their own world in a way that promotes the behaviors you want them to perform.

Procter & Gamble makes a product called Febreze. It is marketed as an odor eliminator. The primary ingredient of the product, beta-cyclodextrin, binds itself to the molecules that cause odors. Once that happens, the molecules cannot attach themselves to the smell receptors in your nose, so the odor is effectively eliminated. It is a fascinating idea.

When the product was first brought to market, customers agreed that it worked. They liked the product, but they did not use it that often. They would buy a bottle, but they would rarely buy a second bottle, because they were not using it often enough to run out of it.

One problem with Febreze was the design of the bottle. Initially, it was sold in spray bottles like those used to sell window cleaners. When people brought the product home, the design of the bottle led them to put it with the rest of the cleaning supplies. For most people, that location is a cabinet under a sink. The only time people reach into that cabinet is when it is time to clean the house, which meant that people would have the opportunity to use Febreze only every once in a while.

Eventually, the bottle was redesigned. Now, it comes in a more attractive cylindrical spray container with printed designs on the outside. It no longer looks like it should be placed in a dark cabinet. Instead, it can be put in a more public place in the house. As a result, people are more likely to put it in a visible location. Because it is more available, it is used more often.

Companies also influence the behavior of other people by easing the shopping experience. The rise of Internet retailers like Amazon have put many book and music stores out of business because it is simply more convenient to order items online than it is to have to go to a brick-and-mortar store.

With the development of handheld devices like e-readers and tablets, commerce has gotten even easier. When the Amazon Kindle was released in 2007, it allowed readers to buy new books directly from the device. Readers could finish one book and immediately buy another.

This ability to create transactions by shaping the environment has reached a peak of efficiency in the tablet gaming market. Video game manufacturers create free or inexpensive games for devices like the iPad and iPhone. These games can be acquired and installed without charge, but several features are available only to those players who pay an additional fee from inside the game application itself. For example, in 2012, the game manufacturer Electronic Arts released a free game based on the animated television show *The Simpsons*. There were several opportunities for players to purchase new features in the game to get access to other characters from the Simpsons universe. In less than a year, the game had generated revenues of over $50 million. Because there was no distance between the desire to make a purchase and the purchase itself, it was easy for players to spend money while playing.

If you can get yourself into people's world, then you can affect their behavior. If you have control over that environment, it is easy to find ways to change the world to make it simpler for people to do what you want them to do. If you do not have control over that environment, you can still use the design of products to affect the way people organize the space around them.

6. Make good behavior cheap and bad behavior expensive. Economists have a concept called an *opportunity cost*, which refers to the idea that whenever you spend time or money on something, that resource is no longer available for anything else. Generally speaking, people are not as sensitive to opportunity costs as they ought to be. They make purchases without really considering all of the other things that they could have done with that money.

At some point, though, almost everyone becomes sensitive to the prices of things. As a result, it is also possible to structure people's economic environments to influence their behavior.

The government influences people's behavior through the tax code. In 1917, the United States amended its tax code to provide a tax deduction for charitable contributions. The idea was that people would not have to declare at least part of the money that they gave to charity each year as income and so it would not be taxed. This deduction makes it attractive for people to give money.

Taxes are also used by the government to push people away from undesirable behaviors. These fees are often called "sin taxes." In 2013, a combination of taxes from New York City and the state of New York added $5.85 to the cost of each pack of cigarettes. As a result, a pack of twenty cigarettes cost well over $10, which doubled the price of cigarettes. Making cigarettes expensive makes them undesirable because that money could be used for other purposes.

The key principle here is that you can use the cost of a product or service to influence behavior by subsidizing behaviors that you would like to see people engage in and by creating penalties for those behaviors you want people to avoid. Price alone will not have a huge influence on people's behavior because they adapt quickly to the current price of goods. But changes in costs can be an effective tool when combined with other methods for affecting people's behavior.

Finally, you can make undesirable behaviors more expensive by increasing the amount of time or effort that it takes to perform an action. Wayne Gray and his colleagues at Rensselaer Polytechnic Institute have found that people are adept at finding the fastest way to accomplish a task. If you slow down the routes toward undesirable behaviors, then you reduce the number of people who do what you do not want them to do. For example, many people will take the elevator in office buildings to go up or down just a few flights of stairs. If you want to encourage more people to walk rather than riding the elevator, then you can slow down the doors of the elevator or the speed of the elevator itself to make the ride less attractive for those who are able to walk.

7. Develop support networks. Customer service has changed in the Internet era. The 1980s witnessed the rise of the big-box retailer. These large stores brought with them economies of scale that allowed them to provide products cheaply and effectively. But, relationships between stores and consumers eroded, because these stores were not part of the larger community. These stores were incredibly successful because they reduced the price of goods. But customer service suffered, which had the potential to cause problems for companies when consumers experienced a problem.

With the rise of the Internet, a number of new avenues for engaging with customers developed along with it. Some have been more successful than others.

Every product now has a website. These websites act like an extended form of advertising, providing information to potential customers. Equally important, they act as a portal for customer service. For products that require occasional software or firmware updates, these sites provide a place for customers to go for downloads. For troubleshooting, most websites provide a list of frequently asked questions and email addresses or Internet forms that allow people to contact customer support. But many of these websites are hard to use and provide a mediocre experience for customers. They also fail to provide information that would support changing people's behavior because they focus on providing information rather than creating neighborhoods.

A more effective means of supporting behavior change comes from bulletin boards for products that require significant engagement on the part of users. These bulletin boards support the development of a community of users who can answer questions and recommend new ways of using products. The bulletin boards can also be monitored by employees of the manufacturer who can answer questions and also find out about problems with the products that can be fixed in future versions.

As one example, professional-caliber recording software is now widely available, allowing musicians and hobbyists to have access, at an affordable price, to the same programs that are used by professional recording studios. The software is incredibly powerful but also difficult to use. The manufacturers of this recording software maintain bulletin board sites that allow users to share recommendations and help new users get acquainted with complex functions of the software. In addition, the companies often add features recommended by users, which engages the user community and encourages them to purchase software updates.

For products and services that do not generate a lot of engagement inherently, it is more effective to create communities around other topics. Parents of school-age children are often deeply involved in their kids' education. A community of other parents facing the same challenges can be a great source of support. Parents want to influence the way their kids study and learn social interactions. They also want to change their own behavior to become more effective parents. Rather than treating some product or service as an outcome, groups like this enable behavior change to be made as part of a larger process, like parenting.

Organizing communities around a process is an efficient way of engaging people to change their behavior. That is the function of groups like Toastmasters International, which aims to help people improve their public speaking skills. Giving talks in public is routinely listed as one of the most stressful events in people's work lives. This anxiety becomes a self-fulfilling prophecy, because the stress of giving a talk hurts people's performance when they get up to give a speech in front of a group. Toastmasters organizes groups of people who get together, give presentations, and give feedback to other members. The atmosphere is professional but relaxed, so the community works to help others get more comfortable with speaking in public. Many people who have been helped by this group continue to attend group meetings to help new members improve their skills. In this way, Toastmasters functions as a source of both mentors and partners in behavior change.

That is ultimately the recipe for a successful support community. Find a process that engages a group of people. Focus on creating a neighborhood around that process. Add experts who can give people good advice to help them achieve their goals. Social relationships are a critical part of behavior change. You can help people who do not have relationships in place already by developing a community to support their efforts at Smart Change and encouraging them to join.

8. Engage in conversations. Another temptation in the modern media environment is to push information at people. Publishing a video on YouTube or creating a website is a way of pushing information. Many social network platforms operate in the same way. Twitter allows users to send off snippets of wisdom and links to a list of followers. Posts on Facebook and LinkedIn also push information out to friends or connections. Even if people choose to comment on a post at these social network sites, it rarely generates much serious discussion.

Although companies have dived into social media as a means of connecting with potential customers, most of them have not seen significant returns on this investment. Because of the importance of real conversation in changing people's behavior, the failure of social media to change people's behavior is not that surprising. Rather than pushing information at people and hoping that their behavior will change, it is important to engage in dialogue with them to help them change their goals and beliefs.

It is a fact that we resist engaging in discussions with people with whom we disagree because it can be uncomfortable to grapple with opinions and beliefs that differ from our own. By definition, conversations about behavior change require people to consider new opportunities that fall outside of their current comfort zone. They are particularly resistant to having conversations like this with strangers, which partially explains why cold sales calls and door-to-door sales are so difficult.

Conversations that change behavior have to grow organically out of relationships that are based on discussion. For this reason, the strategy of trying to engage with people superficially through social media is doomed to failure.

That insight is the basis of grassroots political movements. In 2008 and again in 2012, Barack Obama mounted successful campaigns for the presidency of the United States. A central part of the strategy was to create neighborhoods of supporters. Rather than just pushing advertisements at potential voters, the campaign asked individuals throughout the country to reach out to the people around them and to talk to them. They created templates for events where supporters of the Obama campaign could invite other people over to their homes to have discussions and to watch campaign-related videos.

This campaign did make use of social media, but it did so primarily to further develop local interactions among people. The aim throughout the campaign was to get people talking to each other about the election and to influence their support for candidates. Even more important, this grassroots effort was focused on getting people to act. The Obama campaign realized that simply expressing support for a candidate is not sufficient if people do not get out and vote. As a result, the neighbor relationships that were forged through conversations during the campaign were then used to energize people to go out and vote on Election Day. The campaign was extremely successful at getting out the vote in neighborhoods that traditionally experienced low voter turnout.

The Democrats are not the only ones in politics to learn the lesson of the value of real conversation. In 2009, the Tea Party movement engaged disaffected voters and rallied them around a populist message. Voters who were concerned about tax rates and big government found other like-minded individuals and organized at a local level. These conversations gained momentum in 2009, and in 2010 a number of candidates backed by the Tea Party were swept into office. The Tea Party was also strengthened by conversations among members of the community.

The lesson from these campaigns is that persuasion, influence, and behavior changes are local. Conversations are dialogues. They take place between two (or perhaps a small number) of people. Although you can introduce an idea to a large number of people through mass media, if you really want to change their behavior you have to engage them as individuals.

The Internet can be a valuable tool in creating behavior change, but only if it is used to bring people together for conversations. The aim is not to push information at people but rather to provide opportunities for people to talk in ways that are unrestricted by geography.

The Takeaways

The tools of Smart Change can be used effectively to influence the behavior of other people. However, changing other people's behavior requires addressing the same five elements of the motivational system that are central to changing your own behavior.

To clear the way to affect other people's behavior, you first have to gain their trust. That means that you have to be viewed by other people as acting authentically. Authenticity involves acting in ways that reflect your true goals. When people trust your motives, they engage the Go System in their interactions with you. When they do not trust you, then the Stop System is likely to kick in, which places people in a thinking mind-set rather than a doing mind-set.

Finally, your own leadership is an important way to affect other people's goals. Along the way, you have to consider the kind of feedback you give to people to ensure that you encourage them to focus on processes and keep an incremental mind-set about change. You also need to promote the development of habits and change people's environments to make the desirable behaviors easy to perform and the undesirable behaviors difficult. You need to help people connect with social networks that will allow them to form neighborhoods focused on processes that support Smart Change. These networks create local interactions between people that are supported by conversations.

the smart change journal

Smart Change Journal

Get yourself a blank journal (or download the document at smart changebook.com and go to the "Smart Change" tab if you prefer to work electronically) and respond to the prompts and questions on the following pages to create your own Smart Change Journal. Guided by your own observations, along with the advice I offer in the book, you will be able to develop a plan to achieve your goals and realize your ultimate contribution. For each prompt, I have highlighted the respective chapters from the book that provide more information.

The Big Picture

CHAPTER THREE: OPTIMIZE YOUR GOALS

What do you want to accomplish?

Why is this goal important?

Are you really sure this is the right goal? Why?

Be Specific

CHAPTER THREE: OPTIMIZE YOUR GOALS

On a new page in your journal, create three lists: The actions you can take to advance you toward your goal, the obstacles you may face when you take these actions, and the signs you can use to indicate you have succeeded at the specific actions.

ACTIONS	OBSTACLES	SIGNS

How Can You Fit New Actions into Your Life?

CHAPTER FOUR: TAME THE GO SYSTEM

Engage in a fantasy about how the new actions will fit into your life. Based on that fantasy, develop a specific plan for how to incorporate the new actions into your life. As part of this plan, decide what you are going to do, when and where you will do it, and the people you will need to engage in these actions.

Respond to the following questions for each action you plan to take.

What action are you going to take?

When are you going to do it?

Where will this take place?

How often are you going to need to perform this action?

What aspects of your life will you need to work around to reach the goal?

Whose help do you need?

What resources do you require?

Changes to Your Environment

CHAPTER SIX: MANAGE YOUR ENVIRONMENT

Draw pictures or describe the key environments related to the changes you want to make: home, office, social club, and so on.

What adjustments can you make to your environment to help the changes you want to make become habits?

What changes can you make that will disrupt old habits?

Dealing with others who share your environment:

Who are the people in your home environment?

Who are the people in your work environment?

What other people share your environments?

For each environment, explore how to minimize the disruption of the environment for those people while still supporting your behavior change efforts.

Family, Strangers, Neighbors, and Others

CHAPTER SEVEN: ENGAGE WITH PEOPLE

Family: How can (should) they be involved in your behavior change?

Strangers: Because we tend to be victims of *goal contagion*, identify the Good Strangers and the Bad Strangers who will have an influence on your behavior and on your progress toward your ultimate goal.

GOOD STRANGERS	BAD STRANGERS

Neighbors: Who are the members of your community with whom you engage? How can your neighbors help you avoid temptation?

Mentors: Who are your potential guides and counselors? How can you engage them?

Partners: Who are the people who can share this journey with you? How can you engage them?

Fourteen-Day Habit Diary

CHAPTER SEVEN: ENGAGE WITH PEOPLE

The more you understand about the pattern of your activities, the better able you will be to make changes that bring you into contact with people who will support the goals you want to achieve.

For the next two weeks, at the end of each day respond to the following questions as a way to keep track of your movements and the people with whom you engaged.

Date:

Where did I go?

Did the places I go make it harder for me to keep on track with my goals?

Who helped make change easier or harder?

What other factors made change easy or difficult?

Tracking Your Progress

CHAPTER EIGHT: MAKING CHANGE

Now that you've done the preliminary work, you need to start tracking your progress. Your Smart Change Journal can help. From this point forward, your Smart Change Journal becomes a record of your successes and failures.

At the end of each day, turn to a new page in your journal and take a couple of minutes to write down how your efforts at change are going.

Date:

What were my big successes?

What were the big temptations?

What could I have done better?

ACKNOWLEDGMENTS

I am surrounded by a neighborhood that helps me get things done, and for that I am grateful. My graduate advisers, Dedre Gentner and Doug Medin (to whom this book is dedicated), helped me at the start of my career establish a set of habits that allowed me to get things done. A number of colleagues have helped me learn more about habits and behavior change, including Ayelet Fishbach, Tory Higgins, David Neal, Jamie Pennebaker, Wendy Wood, and Ying Zhang. My graduate students have also been a great source of conversation, leading me to think about these ideas in new ways.

There are many examples of behavior change in practice that have inspired my discussions in this book. The behavioral sciences group at Procter & Gamble has used principles of psychology to examine behavior change at work and at home. They also supported the development of a class called Achieving Peak Performance that got me thinking differently about many of the ideas that ultimately found their way into this book. I appreciate the help of Craig Wynett, Pete Foley, Faye Blum, and Mike Ball. The many students who have taken Achieving Peak Performance have asked questions that have further refined my thinking on the problem of making a contribution.

A number of people talked to me about their efforts to change the behavior of other people. Mike Roizen at the Cleveland Clinic was very generous with his experience. He works tirelessly to make the people around him healthier. I appreciate the time he took to help me understand more of the process that he uses to help others. Isaac Barchas at the Austin Technology Incubator assists new high-tech startups to improve the way they do business. His insights into this process were invaluable.

The staff at the program in the Human Dimensions of Organizations has been important in helping me develop the ideas here in ways that can be put into practice. A big thank-you to Amy Ware and Lauren Lief for their efforts.

Thanks to my sax teacher, Joe Morales. With his impeccable guidance and sage advice, I did achieve my goal of playing in a band.

A number of other people whom I interviewed for the discussions about Joseph Stack and about Alcoholics Anonymous preferred to remain unidentified. I do appreciate their assistance and insight.

Writing a book like this also requires a small community of people. My agent, Giles Anderson, has been a great source of implementation intentions for moving book projects forward. John Duff has an amazing way of making everything I write sound much better. The production staff at Perigee has done another incredible job with the design of this book. My deepest appreciation to everyone.

Finally, while good neighbors are important, good family is even more crucial. I could not have kept focused on this book without the constant support and encouragement of my wife, Leora Orent. My kids, Lucas Markman, 'Eylam Orent Anidjar, and Niv Orent Anidjar, provide a laboratory for thinking about behavior change. Chaviva, the wonder dog, gave me lots of opportunities to take long walks when I needed to think about how to organize my thoughts. My parents, Sondra and Ed Markman, are the best publicists anyone could ever want.

REFERENCES

CHAPTER 1

Cleveland Clinic Program

Roizen, Mike, personal interviews, July 2012.

Brain Size

Perez-Barberia, F. J., and I. J. Gordon. "Gregariousness Increases Brain Size in Ungulates." *Oecologia* 145 (2005): 41–52.

Brains Optimize Energy Consumption

Markman, A. B., and A. R. Otto. "Cognitive Systems Optimize Energy Rather Than Information." *Behavioral and Brain Sciences* 34, no. 4 (2011): 207.

Delay of Gratification

Mischel, W., Y. Shoda, and M. L. Rodriguez. "Delay of Gratification in Children." *Science* 244 (1989): 933–938.

Effort and Accuracy

Gray, W. D., C. R. Sims, W. T. Fu, and M. J. Schoelles. "The Soft Constraints Hypothesis: A Rational Analysis Approach to Resource Allocation for Interactive Behavior." *Psychological Review* 113, no. 3 (2006): 461–482.

Payne, J. W., J. R. Bettman, and E. J. Johnson. *The Adaptive Decision Maker.* New York: Cambridge University Press, 1993.

CHAPTER 2

Cravings

Kassel, J. D., and S. Shiffman. "What Can Hunger Teach Us about Drug

Craving? A Comparative Analysis of the Two Constructs." *Advances in Behavioural Research and Therapy* 14, (1992): 141–167.

Opportunistic Planning

Patalano, A. L., and C. M. Seifert. "Opportunistic Planning: Being Reminded of Pending Goals." *Cognitive Psychology* 34 (1997): 1–36.

Valuation and Devaluation

Brendl, C. M., A. B. Markman, and C. Messner. "Devaluation of Goal-Unrelated Choice Options." *Journal of Consumer Research* 29 (2003): 463–473.

Ego Depletion

Baumeister, R. F., E. Bratslavsky, M. Muraven, and D. M. Tice. "Ego Depletion: Is the Active Self a Limited Resource?" *Journal of Personality and Social Psychology* 74, no. 5 (1998): 1252–1265.

Habits, Willpower, and Stress

Neal, D. T., W. Wood, and A. Drolet. "How Do People Adhere to Goals When Willpower Is Low? The Profits (and Pitfalls) of Strong Habits." *Journal of Personality and Social Psychology* 104, no. 6 (2013): 959–975.

CHAPTER 3

Diets

Dansinger, M. L., J. A. Gleason, J. L. Griffith, et al. "Comparison of the Atkins, Ornish, Weight Watchers, and Zone Diets for Weight Loss and Heart Disease Risk Reduction." *Journal of the American Medical Association* 293, no. 1 (2005): 43–53.

New Year's Resolutions

"Popular New Year's Resolutions." USA.gov. Last revised August 1, 2013. usa.gov/Citizen/Topics/New-Years-Resolutions.shtml.

Positive Thinking: The Good and the Bad

Byrne, Rhonda. *The Secret*. Hillsboro, Ore.: Atria Books, 2006.

Peale, N. V. *The Power of Positive Thinking*. Tokyo: Ishii Press, 2011 (originally published 1982).

Aspect in Language

Slobin, D. I. "From 'Thought and Language' to 'Thinking for Speaking.'" In *Rethinking Linguistic Relativity*, edited by J. J. Gumperz and S. C. Levinson, 70–96. New York: Cambridge University Press, 1996.

Restrained Eating and Diet

Herman, C. P. "Human Eating: Diagnosis and Prognosis." *Neuroscience and Biobehavioral Reviews* 20, no. 1 (1996): 107–111.

CHAPTER 4

Stephen King

Rogak, Lisa. *Haunted Heart: The Life and Times of Stephen King.* New York: Thomas Dunne Books, 2009.

Daily Routines

Daily Routines. dailyroutines.typepad.com.

Designing What You Say for the Audience

Krauss, R. M., and S. R. Fussell. "Perspective-Taking in Communication: Representations of Others' Knowledge in Reference." *Social Cognition* 9, no. 1 (1991): 2–24.

Implementation Intentions

Gollwitzer, P. "Implementation Intentions: Strong Effects of Simple Plans." *American Psychologist* 54 (1999): 493–503.

Dalton, A. N., and S. A. Spiller. "Too Much of a Good Thing: The Benefits of Implementation Intentions Depend on the Number of Goals." *Journal of Consumer Research* 39, no. 3 (2012): 600–614.

Fantasies

Oettingen, G., D. Mayer, A. T. Sevincer, et al. "Mental Contrasting and Goal Commitment: The Mediating Role of Energization." *Personality and Social Psychology Bulletin* 35, no. 5 (2009): 608–622.

Oettingen, G., H.-J. Pak, and K. Schnetter. "Self-Regulation of Goal-Setting: Turning Free Fantasies about the Future into Binding Goals." *Journal of Personality and Social Psychology* 80, no. 5 (2001): 736–753.

Change and Stress

Tennant, C., and G. Andrews. "A Scale to Measure the Stress of Life Events." *Australian and New Zealand Journal of Psychiatry* 10 (1976): 27–32.

Yerkes-Dodson Law

Yerkes, R. M., and J. D. Dodson. "The Relation of Strength of Stimulus to Rapidity of Habit-Formation." *Journal of Comparative Neurology and Psychology* 18 (1908): 459–482.

Arousal and Goals

Carver, C. S., and M. F. Scheier. *On the Self-Regulation of Behavior.* New York: Cambridge University Press, 1998.

Kruglanski, A. W., J. Y. Shah, A. Fishbach, et al. "A Theory of Goal Systems." *Advances in Experimental Social Psychology* 34 (2002): 331–378.

Goal Contagion

Aarts, H., P. M. Gollwitzer, and R. R. Hassin. "Goal Contagion: Perceiving Is for Pursuing." *Journal of Personality and Social Psychology* 87, no. 1 (2004): 23–37.

CHAPTER 5

Shopping While Hungry

Gilbert, D. T., and T. D. Wilson. "Miswanting: Some Problems in the Forecasting of Future Affective States. In *Thinking and Feeling: The Role of Affect in Social Cognition*, edited by J. Forgas, 178–197. New York: Cambridge University Press, 2000.

Drunk Driving Statistics

National Highway Traffic Safety Administration. "2011 Traffic Safety Facts FARS/GES Annual Report." www-nrd.nhtsa.dot.gov/Pubs/811754 AR.pdf.

Psychological Distance

Liberman, N., M. D. Sagristano, and Y. Trope. "The Effect of Temporal Distance on Level of Mental Construal." *Journal of Experimental Social Psychology* 38 (2002): 523–534.

Trope, Y., and N. Liberman. "Temporal Construal." *Psychological Review* 110, no. 3 (2003): 403–421.

Distance and the Strength of Goals

Miller, N. E. "Liberalization of Basic S-R Concepts: Extensions to Conflict Behavior, Motivation, and Social Learning." In *Psychology: A Study of a Science. General and Systematic Formulations, Learning, and Special Processes*, edited by S. Koch, 2:196–292. New York: McGraw Hill, 1959.

The What-the-Hell Effect

Cochran, W., and A. Tesser. "The 'What the hell' Effect: Some Effects of Goal Proximity and Goal Framing on Performance." In *Striving and Feeling: Interactions among Goals, Affect, and Self-Regulation*, edited by L. L. Martin and A. Tesser, 99–120. Mahwah, NJ: Lawrence Erlbaum, 1996.

Protected Values

Baron, J., and M. Spranca. "Protected Values." *Organizational Behavior and Human Decision Processes* 70, no. 1 (1997): 1–16.

Irwin, J. R., and J. S. Spria. "Anomalies in the Values for Consumer Goods with Environmental Attributes." *Journal of Consumer Psychology* 6, no. 4 (1997): 339–363.

Tetlock, P. E., O. V. Kristel, S. B. Elson, et al. "The Psychology of the Unthinkable: Taboo Trade-Offs, Forbidden Base Rates, and Heretical Counterfactuals." *Journal of Personality and Social Psychology* 78, no. 5 (2000): 853–870.

Temptation and the Go System

Fishbach, A., R. S. Friedman, and A. W. Kruglanski. "Leading Us Not into Temptation: Momentary Allurements Elicit Overriding Goal Activation." *Journal of Personality and Social Psychology* 84, no. 2 (2003): 296–309.

Expertise

Ericsson, K. A., and A. C. Lehmann. "Expert and Exceptional Performance: Evidence of Maximal Adaptation to Task Constraints." In *Annual Review of Psychology,* edited by J. T. Spence, J. M. Darley, and D. J. Foss, 47:273–305. Palo Alto, CA: Annual Reviews, 1996.

Ericsson, K. A., and J. Smith, eds. *Toward a General Theory of Expertise: Prospects and Limits.* New York: Cambridge University Press, 1991.

Mind-Sets

Dweck, C. *Mindset.* New York: Random House, 2006.

Mind-Sets and Ego Depletion

Job, V., C. S. Dweck, and G. M. Walton. "Ego Depletion—Is It All in Your Head? Implicit Theories About Willpower Affect Self-Regulation." *Psychological Science* 21, no. 11 (2010): 1686–1693.

CHAPTER 6

Losses Are Painful (Loss Aversion)

Kahneman, D., J. L. Knetsch, and R. H. Thaler. "Anomalies: The Endowment Effect, Loss Aversion and Status Quo Bias." *Journal of Economic Perspectives* 5, no. 1 (1991): 193–206.

Embodied Cognition

Gibson, J. J. *The Ecological Approach to Visual Perception.* Hillsdale, NJ: Lawrence Erlbaum, 1986.

Glenberg, A. M. "What Memory Is For." *Behavioral and Brain Sciences* 20, no. 1 (1997): 1–55.

CHAPTER 7

Types of Relationships

Fiske, A. P. "The Four Elementary Forms of Sociality: Framework for a Unified Theory of Social Relations." *Psychological Review* 99 (1992): 689–723.

Culture

Nisbett, R. E., K. Peng, I. Choi, and A. Norenzayan. "Culture and Systems of Thought: Holistic Versus Analytic Cognition." *Psychological Review* 108, no. 2 (2001): 291–310.

Social Networks

Christakis, N. A., and J. S. Fowler. "The Spread of Obesity in a Large Social Network over 32 Years." *New England Journal of Medicine* 357 (2007): 370–379.

Cohen-Cole, E., and J. M. Fletcher "Is Obesity Contagious? Social Networks vs. Environmental Factors in the Obesity Epidemic." *Journal of Health Economics* 27 (2008): 1382–1387.

Cheating

Mazar, N., O. Amir, and D. Ariely. "The Dishonesty of Honest People: A Theory of Self-Concept Maintenance." *Journal of Marketing Research* 45, no. 6 (2008): 633–644.

Shariff, A. F., and A. Norenzayan. "God Is Watching You: Priming God Concepts Increases Prosocial Behavior in an Anonymous Economic Game." *Psychological Science* 18, no. 9 (2007): 803–809.

Categories

Landau, B., L. B. Smith, and S. Jones. "Syntactic Context and the Shape Bias in Children's and Adults' Lexical Learning." *Journal of Memory and Language* 31 (1992): 807–825.

Malt, B. C. "Category Coherence in Cross-Cultural Perspective." *Cognitive Psychology* 29 (1995): 85–148.

Malt, B. C., S. A. Sloman, S. Gennari, et al. "Knowing Versus Naming: Similarity of the Linguistic Categorization of Artifacts. *Journal of Memory and Language* 40 (1999): 230–262.

Ward, T. B., A. H. Becker, S. D. Hass, and E. Vela. "Attribute Availability and the Shape Bias in Children's Category Generalization." *Cognitive Development* 6 (1991): 143–167.

Categories and Communication

Markman, A. B., and V. S. Makin. "Referential Communication and Category Acquisition." *Journal of Experimental Psychology: General* 127, no. 4 (1998): 331–354.

Joseph Stack

Anonymous, personal interviews with musicians who played at jam sessions with him, February/March 2010.

Alcoholics Anonymous

Anonymous, personal interviews with members of Alcoholics Anonymous, August 1990 through June 1991.

AA General Services. aa.org. The *AA Big Book* and *Twelve Steps and Twelve Traditions* describe the program and the role of mentors within the program. They are available from this site.

CHAPTER 8

Thinking and Doing Mind-Sets

Kruglanski, A. W., E. P. Thompson, E. T. Higgins, et al. "To 'Do the Right Thing' or to 'Just Do It': Locomotion and Assessment As Distinct Self-Regulatory Imperatives." *Journal of Personality and Social Psychology* 79, no. 5 (2000): 793–815.

Kruglanski, A. W., and D. M. Webster "Motivated Closing of the Mind: 'Seizing' and 'Freezing.' " *Psychological Review* 103, no. 2 (1996): 263–283.

Commitment Contracts

Gine, X., D. Karlan, and J. Zinman. "Put your Money Where Your Butt Is: A Commitment Contract for Smoking Cessation." *American Economic Journal: Applied Economics* 2 (2010): 213–235.

Motivational Changes during the Semester

Grimm, L. R., A. B. Markman, and W. T. Maddox. "End-of-Semester Syndrome: How Situational Regulatory Fit Affects Test Performance over an Academic Semester." *Basic and Applied Social Psychology* 34, no. 4 (2012): 376–385.

The Learning Curve

Josephs, R. A., D. H. Silvera, and R. B. Giesler. "The Learning Curve As a

Metacognitive Tool." *Journal of Experimental Psychology: Learning, Memory, and Cognition* 22, no. 2 (1996): 510–524.

Stages of Smart Change

Bonezzi, A., C. M. Brendl, and M. De Angelis. "Stuck in the Middle: The Psychophysics of Goal Pursuit." *Psychological Science* 22, no. 5 (2011): 607–612.

Huang, S., Y. Zhang, and S. M. Broniarczyk. "So Near and Yet So Far: The Mental Representation of Goal Progress." *Journal of Personality and Social Psychology* 103, no. 2 (2012): 225–241.

Zhang, Y., and S. C. Huang. "How Endowed Versus Earned Progress Affects Consumer Goal Commitment and Motivation." *Journal of Consumer Research* 37, no. 4 (2010): 641–654.

Silicon Valley and Boston

Saxenian, A. *Regional Advantage.* Cambridge: Harvard University Press, 1996.

Failure and Self-Compassion

Breines, J. G., and S. Chen Chen. "Self-Compassion Increases Self-Improvement Motivation." *Personality and Social Psychology Bulletin* 38, no. 9 (2012): 1133–1143.

CHAPTER 9

Belief and False Belief

Gilbert, D. T. "How Mental Systems Believe." *American Psychologist* 46, no. 2 (1991): 107–119.

Johnson, H. M., and C. M. Seifert. "Sources of the Continued Influence Effect: When Misinformation in Memory Affects Later Instances." *Journal of Experimental Psychology: Learning, Memory, and Cognition* 20, no. 6 (1994): 1420–1436.

Disgust

Rozin, P., L. Millman, and C. Nemeroff. "Operation of the Laws of Sympathetic Magic in Disgust and Other Domains." *Journal of Personality and Social Psychology* 50, no. 4 (1986): 703–712.

Rozin, P., and C. Nemeroff. "The Laws of Sympathetic Magic: A Psychological Analysis of Similarity and Contagion." In *Cultural Psychology,* edited by J. W. Stigler, R. A. Shweder, and G. Herdt, 205–232. New York: Cambridge University Press, 1990.

Feedback and Stages of Change

Fishbach, A., T. Eyal, and S. R. Finkelstein. "How Positive and Negative Feedback Motivate Goal Pursuit." *Social and Personality Psychology Compass* 4, no. 8 (2010): 517–530.

Motivation and Advancement

Koo, M., and A. Fishbach. "Climbing the Goal Ladder: How Upcoming Actions Increase Level of Aspiration." *Journal of Personality and Social Psychology* 90, no. 1 (2010): 1–13.

Agreeableness and Feedback

Markman, A. *Habits of Leadership*. New York: Perigee Books, 2013.

Checklists and Behavior

Gawande, A. *The Checklist Manifesto*. New York: Metropolitan Books, 2010.

Opportunity Costs

Frederick, S., N. Novemsky, J. Wang, et al. "Opportunity Cost Neglect." *Journal of Consumer Research* 36 (2009): 553–561.

Adapting to the Time It Takes to Perform a Task

Gray, W. D., C. R. Sims, W. T. Fu, and M. J. Schoelles. "The Soft Constraints Hypothesis: A Rational Analysis Approach to Resource Allocation for Interactive Behavior." *Psychological Review* 113, no. 3 (2006): 461–482.

INDEX

ABOUT THE AUTHOR

Art Markman, PhD, is the author of *Smart Change* and *Habits of Leadership*. A graduate of the University of Illinois, he has worked at Northwestern University and Columbia University and has been on the faculty at the University of Texas since 1998 where he is the Annabel Irion Worsham Centennial Professor of Psychology and Marketing.

One of the premier cognitive scientists in the field, Markman has published more than 150 articles and chapters and is the author of the scholarly work *Knowledge Representation*. He has been the editor of *Cognitive Science* since 2006.

As a consultant he has worked with industry, including Proctor & Gamble, for which he developed a number of training programs. His interest in the intersection of business and academia led to the creation of the Human Dimensions of Organizations program at the University of Texas at Austin, which brings the humanities and the social and behavioral sciences to people in business, nonprofits, government, and the military.

Markman has worked with Drs. Mehmet Oz and Michael Roizen on two of their bestselling *You* books and contributes to their social networking website, YouBeauty. He is also on the scientific advisory boards for *The Dr. Phil Show* and *The Dr. Oz Show*.

He has been quoted and interviewed in national and local media from the *New York Times* to *Psychology Today* (for which he also writes and blogs) to *USA Today* as well as many television and radio shows. Markman also contributes to *Fast Company*, the *Huffington Post*, *99U*, and *Harvard Business Review* online.

Follow him on Twitter @artmarkman and at Facebook.com/ArtMarkman. Read more at smartthinkingbook.com.